ADDITIVES ADULTERANTS AND CONTAMINANTS IN BEER

Jeffrey Patton was born in Derby in 1944. He is a qualified brewer and has worked for several breweries from the north to the south of England. He has a particular interest in brewing history and is an enthusiastic member of the Brewery History Society. He has been the editor of *Chronicle* which is the journal of the Heritage Brewery Museum in Burton-on-Trent, since 1985. He now lives in North Devon where he devotes much of his time to writing and publishing.

To Imogen

Is it good?

ADDITIVES

ADULTERANTS AND CONTAMINANTS

IN BEER

JEFFREY PATTON

PATTON PUBLICATIONS

First published in 1989
by
Patton Publications
Orchard House
Swimbridge
Barnstaple
North Devon
EX32 OPL

BRITISH LIBRARY CATALOGUING IN PUBLICATION DATA

Patton, Jeffrey
Additives, adulterants and contaminants in beer.
1. Great Britain. Brewing industries, history
I. Title
338.4'76633'0941

Set in 10 on 12 Times Roman
Copy preparation by Patton Publications
Printed and bound in Great Britain
by
BPCC Wheatons Ltd, Exeter

ISBN 1 872426 00 X

'Some Coopers attempt to extend their Art, so far as to add Strength to the Beer; but let it be remembered, that the Principal Constituent Parts of Beer, *should be* Malt and Hop; when strength is given to the liquor by any other means, its Nature is altered, and it is no more BEER that we drink.'

Combrune, M., *Theory and Practice of Brewing*, 1762.

Note: The coopers referred to were not the makers of casks but tasters employed by the large porter breweries who saw that the beer blended in the brewery was of the correct taste and quality. Abroad Coopers advised licencees and attended to beer quality in the trade.

CONTENTS

ILLUSTRATIONS

PREFACE

There seems to have been much more concern in recent years about what we eat. This concern has always existed to some extent but recently the consumer's interest has been stimulated much more by pressure groups, product labelling, books about food additives and a greater emphasis on the subject by the press and the television. We are now certainly all aware that food technology is very advanced and that today it is almost impossible to avoid food additives. These may be regarded by some as a boon and by some as a bane, depending on one's viewpoint. However, food adulteration is an area which unquestionably seems to have improved for the consumer whereas contamination has only changed. A hundred years ago our concern may have been over finding some animal life or their remains in our food, but now our concern may be for less visible evidence such as pesticide residues which are the legacy of having the former problem resolved.

Although beer is consumed in great quantities, passes though our gastro-enteric tract where it is digested and absorbed to provide nutritional benefit, it is often ignored as a food. Many diet studies ignore the contributions made by beer as do many books written about food additives. It is hoped that this book will provide the person who includes some beer in his diet, with enough information for him to asses the contribution it makes, whether good or bad.

Taking a broader view, this book also explains why brewers may need to use additives and closely examines the contaminant problems of beer. Beer contaminants are unique, and perhaps surprisingly, some were entirely unknown to brewers twenty years ago. Some are of mans' own making but others have always been there and have only recently come to light through the results of scientific research.

The brewing industry in Britain is one of the oldest and like others has undergone much change during its long history. Unlike Henry Ford, the author sets great value by the lessons of history. The past is the only true reality which we can use to evaluate the present. It is hoped that at best the inclusion of historical facts will stimulate the reader into looking at the subject a little differently, but that at least it will provide an interesting insight into the past. As an example, many readers who believe things get worse as time goes on, will be surprised to read of the problems associated with the adulteration of beer in the last century, a fascinating aspect of brewing history.

Other topics covered by the book include: a section on the many different herbs which have been used for various purposes in brewing over the centuries, a section on food legislation relating to beer, a section on the more unusual ingredients which have been used in exotic beers and another on the slightly more controversial matter of 'pure' beer.

The author wishes to thank everyone who has been of assistance to him in the course of his research: those who have responded to requests for information, numerous librarians and archivists, and the authors of many books and articles most of which are included in the bibliography.

J D Patton

Swimbridge
September 1989

ABBREVIATIONS

ADI	Acceptable Daily Intake (mg/kg body weight)
ATNC	Apparent Total N-Nitroso Compounds
BCMA	British Caramel Manufacturers' Association
BIBRA	British Industrial Biological Research Association
BRF	Brewing Research Foundation
brl	Barrel (36 gallons)
COC	Committee on Carcinogenicity
COM	Committee on Mutagenicity
COT	Committee on the Toxicity of Chemicals in Food, Consumer Products and the Environment
dl	Decilitre (a tenth of a litre)
DEHA	Di-2-ethylhexyl adipate
DEHP	Di-2-ethylhexyl phthalate
DIOP	Di-isooctyl phthalate
EEC	European Economic Community
FAC	Food Advisory Committee (of MAFF)
FACC	Food Additives and Contaminants Committee (of MAFF)
FAO	Food and Agriculture Organisation (of United Nations)
FDA	Food and Drug Administration (U.S.A.)
gm	Gram
hl	Hectolitre (a hundred litres)
ITCA	International Technical Caramel Association
JECFA	Joint Expert Committee on Food Additives
kg	Kilogram (a thousand grams)
MAFF	Ministry of Agriculture, Food and Fisheries
mg	Milligram (a thousandth of a gram)
NDMA	N-nitrosodimethylamine
OG	Original gravity
ppm	Parts per million (Same as mg/kg)
ppb	Parts per billion
P	Plato
PET	Polyethylene terephthalate
PVPP	Polyvinylpolypyrrolidone
S.I.	Statutory Instrument
μg	Microgram (a millionth of a gram)
WHO	World Health Organisation

INTRODUCTION

The subjects of this book are all the substances which can be used to make beer or are incorporated into beer other than the main ingredients: malted barley, hops, water and yeast. They have been classified into three groups: additives, adulterants and contaminants. The distinctions made between the three are that additives are legal ingredients which are used in producing wholesome beer, adulterants are illegal ingredients or substances added after brewing, and contaminants are prohibited or undesirable substances which are found in beer through accident or neglect. The subject is a little more complex than one would first suppose, so it is vital for complete clarity that the three groups are more fully defined. The definitions given here are only intended to explain how the words are used in this book, the aim of which is to give information on the history, nature and use or occurrence of the substances discussed. It must be appreciated that other authors writing from a different viewpoint or for a different purpose may use a different system of classification and different definitions of the substances involved.

ADDITIVES

The proverb 'One man's meat is another man's poison' could be rewritten 'One man's additive is another man's poison'. It is not a straightforward task to give a definition of an additive which suits every circumstance in which the word is used. The use of additives in food is controlled by law, so there are a number of definitions which are worded in legalistic language in order to leave no ambiguity of meaning. However, they are usually a little limited in their general usefulness. There are a number of properties which additives can possess, but they do not necessarily apply to every additive. The only implicit property which can be said to be possessed by every beer additive with certainty, is that it is added to beer. Rather than give a complex definition covering every circumstance, it is perhaps more helpful to give a list of properties which are generally applicable to most additives:

1. They are not normally consumed as a foodstuff by themselves.
2. They are not basic food ingredients. Although they may improve the product or make its manufacture easier, they are not necessarily vital ingredients.
3. They are of little or no nutritive value.

4. They are added for a technological (including organoleptic) purpose during manufacture, processing, preparation or treatment of the food and affect its character or nature.
5. They become incorporated in the product.
6. Their rates of usage are usually at a low level.
7. Their use is usually the subject of official scrutiny, testing and legislation.
8. They are not often raw materials but manufactured products of other industries. This is perhaps why additives are labelled chemical or artificial in many people's minds.
9. Their effect is often important and seems to outweigh their level of usage.

Whether a substance is a permitted additive depends entirely on the laws of the country in which the food is sold, so an additive in one country can be an adulterant in another country. Levels of use are also controlled, so an additive used above the permitted concentration can become an adulterant. The way in which additive legislation is made differs from one country to another. Permitted additives may be listed on a positive list or prohibited additives may be listed on a negative list. In Britain additives must be listed on most food labels. One of the exceptions is alcoholic drinks where the alcoholic strength is more than 1.2% by volume. This includes most beers.

PROCESSING AIDS

These are substances or materials, excluding the manufacturing equipment, which are used to produce a desirable physical effect during manufacture, processing or treatment of a food. When the technological purpose has been fulfilled they are removed in a later part of the manufacturing process. Sometimes a little of the processing aid may be left behind in the food as it is often difficult to remove it completely. Specified amounts of residual processing aid are usually permitted but above these levels they are regarded as contaminants. Labelling regulations in the U.K. do not require processing aids to be listed on food labels.

ADJUNCTS

These are brewing materials which are a source of extract and can be used to replace or supplement malt. They have a nutritive value and are used in amounts of up to 50% of the total grist and occasionally in even higher amounts. In countries where their use is permitted they are regarded as wholesome brewing materials, but in countries where there are restrictions they can be regarded as adulterants. The reason why they have been included in this book is because there has been some controversy over their use in beer in recent years. Some people regard them as being undesirable additives in beer even though they accept that they are wholesome substances which are legally permitted.

ADULTERANTS

These are substances which could be described as non-permitted additives. They are not permitted either because legislators see no need for them or because they are regarded as harmful. They are used mainly because they offer some advantage to the adulterator, which is usually of a financial nature. Since they are used to deceive the consumer and because their presence is easy to detect their use in brewing in the U.K. was largely eradicated a hundred years ago. However, adulteration of beer by some retailers is still carried out.

CONTAMINANTS

These substances when found in beer serve no useful purpose to anyone. They occur by accident or by negligence on the part of the brewer or retailer. They are usually present in very small amounts which often leads to their presence remaining undetected. They can be transferred to beer from contaminated brewing materials or from the materials of which the brewing plant or equipment is made. As mentioned, they may also be residual amounts of processing aid or additives used over the recommended limits. Many contaminants are harmless but a few, like lead and arsenic, are highly toxic and others can have long-term cumulative adverse effects.

Beer may also become contaminated with micro-organisms which grow in beer causing spoilage, but it is not an aspect of the contamination of beer covered by this book.

CHAPTER 1

EARLY HISTORY

Beer is known to have been brewed in Britain by the Celts, well before the time of the Roman occupation. In early times brewing was essentially a domestic process like baking, spinning and weaving. Such everyday tasks and events were taken as a matter of course, so not much was written about them. Skills were passed down from person to person by practical demonstration rather than by written instruction. Think of where you would look today for written instructions on how to make a cup of tea if you had never been shown what to do. You will have difficulty to find anything in a cookery book or even on the back of a packet of tea. It was not until the start of industrialisation that details of the brewing process and the ingredients used were recorded. This was so that masters of the art and science could pass the benefits of their knowledge to others. Despite the situation, there are some facts on the subject which can be gleaned from early literature.

ADDITIVES

HERBS AND SPICES

Herbs were occasionally added to beer for medicinal purposes. It was discovered by the Egyptians that some of the essential substances in herbs would dissolve in the alcohol found in beer, whereas they would not dissolve in water. Beer was also safe to drink as it was free of disease-carrying bacteria and covered up the taste of some of the more unpleasant herbs. Herbalists such as Gerard writing at a later date often used wine for a similar purpose. Betony, a herb which relieves coughs, is an example of one of the more popular medicinal herbs used in Britain in early times.

Various herbs and spices were also used as flavourings, the hop being only one of many until its superiority became established at the end of the Mediaeval period. Spiced beer was a well-known type in early times and because of the expensiveness of spices it was a great luxury. This particular type of beer is described in more detail in Chapter 8. Native herbs which could be gathered from the wild were used in the more common beers.

There are few details available of the herbs used in early Mediaeval times, though an old Irish poem of the twelfth century indicates that the Saxons used a bitter herb for flavouring. It certainly would not have been the hop and most likely was the bog bean.

> The Saxon ale of bitterness
> Is drank with pleasure about Inber in Rig,
> About the land of the Cruithni, about Gergin,
> Red ale, like wine, are freely drank.

Ling or ling-heather is another herb known to have been used in the period. Its first use may have been even earlier but the first recorded use is by the Danes in Ireland in the ninth century. It has a long history of use and the details are given together in Chapter 8.

A herb which appears to have been used widely in the Mediaeval period was alecost and gave beer a spicy, balsamic flavour which was popular at the time. Myrica gale was another well-liked herbal flavouring of the period and its use continued for some time afterwards. It was originally used by the monks and the beer produced with it was known as gale ale or gale beer. The last area in which it was produced was Yorkshire. It was also an ingredient of gruit.

GARDNER HARVESTING HERBS 1477

GRUIT

Gruit was a mixture of herbs and other substances which were used for flavouring beer in Mediaeval times. Monks originated the formulation, which was widely used throughout Northern Europe, particularly in England, the Netherlands and Germany. The communications between the brotherhoods of monks explains its widespread use despite the fact that its formulation was kept secret. Most of the information on gruit comes from German documentation. The use of gruit in Germany was well organised and well documented because it was used as a source of revenue for the Church or the State. Various combinations of herbs were used, but all were basically similar and were known by several similar names besides gruit, including grut, gruz, grug and gruys. The earliest reference to this substance is from the writings of the Abbess Hildegarde of St Ruprechtsberg, near Bingen, who in 1079, in her Physica Sacia, mentions a vat of beer being boiled with grug. She also mentions the use of myrtle berries, ash leaves and hops.

The exact combination of the herbs used in gruit was a secret known only to a privileged few, but as the herbs were the accepted flavour of the time people had to buy the gruit from the privileged manufacturers. This privilege was sold to the manufacturers to provide an income known as the gruitgeld. It could also be a tax paid by the people who bought the manufactured gruit. The authorities protected their position against the rising popularity of hops by prohibiting their use with severe penalties for any infringements.

The major recipient of the gruitgeld was the Archbishop of Cologne who controlled the privileges over a large area including several towns and cities. Naturally the municipal governments had their eye on the gruitgeld and fought the Church for the right to control it. In 1278 the town of Munster obtained the gruit monopoly and the gruit house from the Archbishop of Cologne by purchasing a third of it, but also had to pay 40 marks in perpetuity to the Cathedral Chapter. Eventually in 1663 the town rid itself of this obligation by a further payment. Cologne did not obtain the gruitgeld until 1500, but by this time the brewers had ceased using gruit and presumably the only people using it were private citizens.

The three major ingredients of gruit were:

1. *Myrica gale*, also known as sweet gale, pors, porze or porst.
2. *Ledum palustre*, also known as marsh or wild rosemary.
3. *Achillea millefolium*, also known as milfoil or yarrow.

The composition, however, did vary from district to district. An inventory in the accounts of the Archbishop of Cologne, in 1393, not only mentions *Myrica gale* and *Ledum palustre* but also juniper berries, ginger, caraway seed, aniseed and several other herbs and spices. Along the Rhine, especially in Cologne, *Myrica gale* was the main ingredient, whilst in Westphalia *Ledum palustre* was more popular as it grew in profusion in the district.

It was usual to mix the herbs with ground malt in order to disguise them. The gruit malt was then mixed with the ordinary malt at the rate of 10 pounds per quarter (336 pounds).

Hops had been used alongside gruit and their earliest recorded use is in the same passage written by Abbess Hildegarde which records the earliest use of gruit. Despite the legislation brought in against them, hops slowly began to replace gruit as a flavouring from the fourteenth century, until in the sixteenth century very little gruit was in use. Eventually legislation was brought in prohibiting the use of gruit. Provincial laws were passed in Bavaria in 1553 and 1616. Holstein, in 1623, expressly forbade the use of bog myrtle, which with other herbs was described as an unhealthy material. In Brunswick Luneburg, in 1723, bog myrtle was banned as imparting a dangerous potency after its use had continued despite earlier warnings that it was putting the lives of the subjects in peril.

In Britain one of the few references to gruit was in the registry of the priory at Worcester where it is mentioned that four quarters of 'growte' were mixed with ten quarters of mixed barley and oat malt in a brew. Strangely enough, it was in England and in Devon in particular that gruit was last used, probably because the force of the legislation in Germany soon eradicated its use over there. On the Continent there was fierce competition between the different cities and municipal authorities, and through local legislation they exercised a greater control over trade than British cities and towns. As far as beer was concerned they often controlled how it was made, when it was made and who made it. The character of the beer was thus determined by the town in which it was brewed rather than by individual brewers. Though certainly good for the general prosperity of the individual towns, it did stifle individuality at a local level. This situation did not exist in Britain and would have been the reason why the production of a gruit beer was able to linger on in Devon until the nineteenth century.

DEVON WHITE BEER

The gruit beer of Devon was known as Devon White Beer. The earliest documentation giving any details of its nature dates from the eighteenth century. Even then it was recorded as being something of an oddity and was referred to as a beer of antiquity. The essential ingredient of the beer was grout, also known as ripening. Like the German gruit, the exact composition of it was a secret known only to a few. In the early eighteenth century it was said that there were only 'two or three masters of the secret' and that they sold it as yeast was sold. A great lump or ball, which obviously contained yeast, was enough to ferment four or five steens (a steen is about five gallons) of wort. It also contained a mixture of herbs which like gruit were disguised by ground malt. This explained the fact that it thickened the ale to give it a white appearance, from which its name was derived. Egg white was believed to have been another ingredient and bound the other ingredients together in the lump.

By the early nineteenth century the secret was believed to be in the possession of only one family.

Charles Vancouver, in 1808, records the following details of its history and a somewhat modified method of production:

> The brewing of a liquor called white ale, is almost exclusively confined to the neighbourhood of Kingsbridge: its preparation, as far as could be learnt by the Surveyor, is 20 gallons of malt mashed with the same quantity of boiling [sic] water; after standing the usual time, the wort is drawn off, when six eggs, four pounds of flour, a quarter of a pound of salt and a quart of grout, are beat up together and mixed with this quantity of wort, which, after standing twelve hours is put into a cask and is ready for use the following day. This beverage produces a very intoxicating quality, and is much admired by those who drink not to quench thirst only. . . .
>
> That this liquor is of considerable antiquity is plain, from the terrier [a document not a dog] of the advowson of Dodbrook, and which expressly calls for the tithe of white ale. The present worthy incumbent commutes this claim, for half a guinea annually from each house in the parish where this ancient beverage is retailed

It bears a similarity to the situation in Cologne where the Gruitgeld provided the Church with an income source.

The antiquity of the ale is confirmed by Rev. Richard Polwhele writing in his *History of Devonshire* 1793–1806, in which he states that the people of Kingsbridge have made it from time immemorial and indicates that it was at least being brewed in the time of Henry VIII. However, the *London and Country Brewer*, published in 1742, records that it was first brewed only sixty years earlier. An explanation of these contradicting facts is offered. From the eighteenth and nineteenth century writings it is clear that the grout is mixed with the yeast and was used to start the fermentation, whereas the gruit in German beers was boiled with the wort in the same way as hops are. The inclusion of the grout herbs with the flour and yeast may have been a development originating sixty years prior to 1742 and the date of origin of white beer as it was known in the eighteenth and nineteenth centuries.

Testimony to the worth of white ale can be summed up in one sentence from the *London and Country Brewer*:

> Now this white ale being thus fermented into such a gross body, becomes a sort of chyle ready prepared for digestion in the stomach, and yet so liquid to pass the several secretory ducts in the animal system soon enough to give room for new supplies of this pleasant tipple, even at one common sitting in a public house: for though this drink is not so thin and clear as the brown sorts, yet, by its new, lubricous, slippery parts, it is soon discharged out of the stomach; and not withstanding such evacuations, it leaves a very nutritious quality behind it in the body, that brings it under a just reputation for preventing and recovering those who are not too far gone in consumptions; and therefore would be of extraordinary service to labouring people: in short, this famous liquor is of such a salubrious nature as renders it a most agreeable drink both to the sedentary and active person; which plainly shews the transcendent quality peculiar to this white ale, beyond all other malt liquors whatsoever ... and at the same time administers to the body such medicinal assistance, that no other malt drink, wine, nor any other potable liquor now in use, as I know of, comes up to it, not only for the aforesaid intention, but it is likewise usually prescribed by physicians, as a remedy in the cholic and gravel by its lubricating, diuretic nature; and it is the best liquor in the world for a wet nurse to drink for increasing a most nourishing milk.

Alas, this beer is no longer brewed so we have no chance to confirm the qualities attributed to it; and indeed even at the end of the nineteenth century the nature of the beer had been so much diminished that it was recorded as being made in Tavistock from common ale with flour and yeast added.

ADULTERANTS

ALE CONNERS

The first indication that there was ever concern over the quality of beer was in 1377, when the first ale conner was appointed in London. His duties are set out in the *Liber Albus* compiled in 1419. Part of these read:

> And that you, so soon as you shall be required to taste any ale of a brewer or brewster, shall be ready to do the same; and in case that it be less good than it used to be before this cry, you, by assent of your Alderman, shall set a reasonable price thereon, according to your discretion; and if anyone shall

> afterwards sell the same above the said price, unto your Alderman you shall certify the same. And that for gift, promise, knowledge, hate, or other cause whatsoever, no brewer, brewster, huckster, cook, or pie-baker, who acts against any one of the points aforesaid, you shall conceal, spare, or tortuously aggrieve; nor when you are required to taste ale, shall absent yourself without reasonable cause and true, but all things which unto your office pertains to do, you shall well and lawfully do. So God you help and the Saints.

The duties clearly indicate that his purpose was to ensure that beer was of good quality and to penalise the brewers of poor beer by reducing the price. His position was obviously a deterrent to anyone contemplating adulteration.

In order to indicate to the ale conner that there was ale to be tasted, the brewer or alewife erected a pole or ale stake outside their premises. It is recorded in 1462 that a brewer, John Lautroppe, had brewed three times but only put out his ale stake once and had thus exposed the public to the danger of being supplied with beer which was 'not sufficiently mighty of the corn, or wholesome for man's body'.

During the next century similar officials known merely as ale tasters were appointed in towns and cities outside the capital. An ale taster in Worcester had the following instructions:

> You shall resort to every brewer's house within this city on their tunning day and there to taste their ale, whether it be good and wholesome for man's body, and whether they make it from time to time according to the prices fixed. So help you God.

A report of the Leet at Coventry in 1439 records details of the position of ale taster; and in 1452 Stratford-on-Avon is recorded as having two ale tasters. It is interesting to note that John Shakespeare, father of the famous William, was appointed an ale taster in Stratford-on-Avon in 1556. The University of Oxford even appointed an ale taster; he was licensed to his office by the Vice-Chancellor and went around all the colleges, which then had their own breweries. He was awarded a fee of one gallon of strong ale and two gallons of small wort for each brew passed. In the eighteenth century the services of the ale taster became less necessary with the advent of the common brewer and development of control by H. M. Excise. They took on other functions or else the appointment became one of name only.

The brewers were expected to show a degree of self regulation as in the seventh year of the reign of Henry IV (1405–1406), an ordinance was made in respect of the Mystery (Trade) of Free Brewers of the City of London in which a constitution was ordained for them by the City

Fathers. It, amongst other things, required them to supervise those who made and supplied ale and to see that 'good, abel and seyn [sound] ale was brewed'.

PUNISHMENT

Punishment for the adulteration of beer and breaking other laws in connection with beer has varied over the years, but it tended to be more severe in olden times. A typical example of punishment in the thirteenth century was that enforced in Danzig, Germany, for the production of poor beer. There the offender was transferred to the dung hill for his crime. This, however, was mild compared with the punishment for avoiding payment of tax on beer in Aix-la-Chapelle, France. There offenders had their right hand chopped off.

In London in the thirteenth century a brewster who contravened the Assize of Bread and Ale, which regulated the prices of the two commodities, was fined, but if she persisted she was subjected to corporal punishment in the tumbrel. This was the name given to a farm dump cart or wagon used for manure; but for the purpose of punishment it was used to parade the offender through the streets. The punishments changed

THE TUMBREL

and in the reign of Edward III (1327–1375), for the first offence the ale was forfeited, for the second the guilty person was thrown out of their trade and for the third they were thrown out of the city. In the reign of Henry V (1413–1422), for the first two offences they were fined but for the third offence corporal punishment was administered in the tumbrel or pillory. In some places such as Scotland the cucking stool was used. It was stated:

> Gif she makes gude ail, that is sufficient. Bot gif she makes euel ail, contrair to the use and consuetude of the burg, and is convict thereof, she shall pay ane unlaw of aucht shillinges, or sal suffer the justice of the brurg, that is, she sall be put upon the Cock-stule, and the aill sall be distributed to the pure folke.

It was also the practice of the city of Chester to punish dishonest ale wives with the cucking stool in Saxon times. This should not be confused with the ducking stool which was a later invention of the early seventeenth century. The cucking stool was merely a seat on which the offender was paraded through the streets with bare head and feet, though they were also sometimes ducked in water. It was obviously something like the tumbrel if not the same thing. A ducking stool on the other hand was a seesaw-like arrangement with a chair on one end which could be lowered into a pond or river. The person being punished was strapped into the chair.

THE ADULTERATORS

In the later Mediaeval times we get a more positive picture showing that ale was suffering a higher degree of adulteration and the ale wife was often indicated as being the main villain. In the Chester Miracle Plays of the fourteenth century, the Chester Tapstere speaks the following words which record her evil doings:

> Some time I was a tavenere,
> A gentel gossepp, and a tapstere
> Of wine and ale, a trusty brewer,
> Which woe hath me bewrought.
> Of cannes I kept no true measure,
> My cuppes I solde at my pleasure,
> Deceavinge many a creature,
> Tho' my ale were nought.

The ale wife is then carried off into Hell's mouth by the attendant demons and the play closes. The illustration shown is based on the play and taken from a miserere seat in Ludlow Church. A demon is about to cast the deceitful ale wife into Hell's mouth. She carries her gay head attire and her false measure whilst another demon reads her offences and a third gives musical accompaniment on the pipes.

THE FATE OF THE ALEWIFE

In 1434 there were problems with the quality of beer in Oxford and it was recorded 'how great evils arise both to the clerks and to the townsmen of the city of Oxford owing to the negligence and the dishonesty of the brewers of ale'. The brewers were all brought together by the Commissary in the Church of the Blessed Virgin Mary, where each one was made to solemnly swear on the holy evangelists that he would thereafter 'brew ale that was good and wholesome as far as his ability and human frailty permitted him'. The University authorities laid down strict instructions on how beer was to be brewed and as already mentioned they had their own ale taster in order to see that these bylaws were enforced. Even so, it became necessary in 1449 for the stewards and manciples of Canterbury College to lodge a complaint against nine brewers. They were alleged to have brewed weak and unwholesome ale which was not properly brewed and not worth the price charged for it. The brewers were again made to swear that they would brew in a wholesome manner. They agreed to attend to the following points in their brewing: that they would continue to heat the water over the fire as long as it continued to emit froth, that they would skim off the froth and that they would not send out new ale until it had had long enough for the dregs to settle out. Incidents continued to occur and in 1464 it was recorded that the Commissary ordered a brewer John Janyn to refund eight pence to Anisia Barbour to whom he had sold a cask of beer for twenty pence. The judgement said: 'In our opinion and that of others who have tasted it, it is not worth more than twelve pence.'

THE NATURE OF ADULTERATION

There are few details of the nature of the adulteration that was carried out, but certainly one of the most popular practices was dilution. It has always been popular, as if carried out to a modest degree it is almost undetectable. This was especially so in times when there was no chemical analysis and detection relied entirely on the tasting ability of the ale taster.

The purpose of dilution was to increase the amount of profit on a cask of beer but another reason for the adulteration of beer was to induce the drinker to drink more of it. Harrison gives the following account in the sixteenth century when writing on the subject of fraudulent ale wives:

> Such sleights have they for the utterance of this drink [ale] that they will mire it with resin and salt, but if you heat a knife red hot, and quench it in the ale, so near the bottom of the pot as you can put it, you shall see the rosen come forth hanging on the knife. As the force of salt, is well known by the effect; for the more the drinker tipleth, the more he may, and so dooth he carry oft a drie drunken noll to bed with him, except his luck be the better.

Jacob Theodor von Bergzabren, who wrote under the name of Tabernaemontanus, wrote a Herbal in 1588 in which he mentions some of the malpractices carried out by German brewers. Some used willow leaves instead of hops and others used chimney soot to give the beer a strong brick-red colour. He wrote:

> Not only does such a beer not taste good, but it spoils the blood, burns it up, causes great thirst, horrible red faces, also leprosy, swelling of the body, injury to the head and all internal parts of the intestines.

He also mentioned the use of Indian beans (*Cocculus indicus*) which were in popular use as an adulterant in much later times, at the beginning of the nineteenth century.

CONTAMINANTS

THE PURITY OF BEER

In Mediaeval times the people who were brewing on the largest scale were the monks, who not only brewed for their own domestic consumption and for those associated with the monastery, but also supplied travellers with refreshment. This was considered a Christian duty, although it was sometimes abused particularly by the nobility. The monks brewed different types of beer for the different classes of people but it can be assumed that it was all wholesome. It is said that the origin of the X marks on the heads of casks of beer came from the practice of the monks to mark the casks with a cross to show that the beer was of sound quality and fit to drink. The cross was a sign that this guarantee was 'sworn on the cross'. Sadly most of our British monasteries were shut down by Henry VIII along with their breweries, but there are still a number with breweries on the continent of Europe. They have a reputation for excellence so it is probable that the monasteries set a standard for others to aspire to.

CHECKING THE ALE

The only other people brewing on anything greater than a domestic scale, though not very much greater, were the ale wives who supplied beer for the local population in their ale houses. One of the earliest comments on the quality of ale was from Henry of Avranches writing in the thirteenth century on the subject of Modbury ale from Devon. The unexpurgated version, which was censored when quoted in Victorian writings, reads:

> Of this strong drink, much like to Stygian lake,
> Most term it ale, I know not what to make.
> Folk drink it thick, and piss it out full thin.
> Much dregs therefore must needs remain within.

The offending word was usually left out or left in the Latin in which it was originally written.

It is of course not known whether the solid inclusions were something put into the beer or if they were the result of spoilage. Technology was rudimentary compared with what is known today and substances which make beer bright had not been discovered. Certainly an attempt to remove the yeast would have been made by allowing it to settle out unaided, but this is no guarantee of a bright beer.

Equipment and containers are difficult enough to sterilise these days with modern sterilants, so it is likely that most equipment, which was often made of wood, would be impregnated with a natural microflora of wild yeasts and bacteria. These would give the beer fruity and acidic flavours, which are found in Belgium Lambic beers produced today; they are produced by allowing fermentation to start spontaneously by the natural microflora living in the brewery. Today the bitterness of hops is used to balance the natural sweetness of beer but as they were not introduced to Britain until late Mediaeval times it was balanced to some extent by the acidity of lactic acid. This was produced by lactic acid bacteria which we would now regard as spoilage micro-organisms but then coexisted with the yeast and be one of the main indigenous bacteria in the brewery. This situation still exists today with native African beers, where naturally occurring lactic acid bacteria are allowed to produce a high level of lactic acid in the beer.

BAD BREWING PRACTICE

Beer did not necessarily become adulterated in early times through a direct deliberate act, but became adulterated through a lack of knowledge of the art or by the plain use of bad practices. An example of bad brewing practice is found in the poem 'The Tunnyng of Elynour Rummyng' by John Skelton, the Poet Laureate, written about 1517. The ale wife Elynour Rummynge was a real person who has been identified as living in Leatherhead in 1525. Skelton wrote of her:

Her lothely lere
Is nothynge clere,
But ugly of chere,
Droupy and drowsy,
Scurvy and lowsy;
Her face all bowsy,
Comely crynklyd,
Woundersly wrynklyd,
Like a rost pygges eare
Brystled with here.

Her lewde lyppes twayne,
They slaver, men sayne,
Lyke a ropy rayne,
A gummy glayre.
She is ugly fayre:
Her nose somdele hoked
And camously croked,
Never stoppynge
But ever droppynge;
Her skynnne lose and slacke,
Greuyned lyke a sacke;
With a croked backe.

ELEANOR RUMMYNG

and of her brewing methods he wrote (with translation by the author):

The hennes ron in the mashfat;
For they go to roust
Streyght over the ale joust,

And donge, whan it comes,
In the ale tunnes.
Than Elynour taketh
The mash bolle and shaketh
The hennes donge awaye,
And skommeth it into a tray
Whereas the yeest is,
With her maungy fystis.
And somtyme she blennes
The donge of her hennes
And the ale togyder,
And sayth, gossyp, come hyder,

This ale shal be thyker
And floure the more quycker;
For, I may tell you,
I lerned it of a Jewe
Whan I began to brewe,
And I have found it trew.
Drinke now whyle it is new;

The hens run in the mashtun;
For they go to roost
Straight over the fermenting vessel,
And dung, when it comes,
In the ale runs
Then Elynour takes
The mash bowl and shakes
The hens dung away,
And skims it into a tray
In which the yeast is,
With her maungy hands.
And sometimes she blends
The dung of her hens
And the ale together,
And says, good friend, come hither,
This ale shall be thicker
And froth more quicker;
For I may tell you,
I learnt it from a Jew
When I began to brew,
And I have found it true.
Drink while it is fresh;

And ye may it broke, It shall make you loke Yonger than ye be Yeres two or thre, For ye may prove it by me. Behold, she sayd, and se How bright I am of ble!	And if you can withstand it, It will make you look Younger than you be Years two or three, For you may prove it by me. Behold, she said, and see How bright I am of complexion!

Other sources of the time show that serious faults existed elsewhere as well.

MALT

The following regulation taken from a London Letter Book of the fifteenth century indicates that there were shortcomings in the practice of malting at that time. The malt used was to be:

> ... clene swete and drye and wele made, not capped in the sakkes, nor raw dried malte, dank or wete malt or made of mowe brent barley [barley burnt or overheated in the mow or stack], belyed [i.e. swollen] malte, edgrove [i.e. germinating] malte, acrespired [i.e. sprouting at both ends] malte, wyvell eten [weevil eaten] malt or medled.

Beer was often smoky in taste because the malt had been kilned with a smoky fuel. Andrew Borde's verse on Cornish ale written in dialect, as well as observing the thickness of the ale, indicates that it was also smoky.

> Iche cam a Cornyshe man, ale che can brew;
>   It wyll make one to kack, also to spew;
> It is dycke and smoky, and also it is dyn;
>   It is lyke wash as pygges had wrestled dryn.

Andrew Borde was a physician and traveller as well as being a man of wit and wisdom. He wrote accounts of the food and drink of his time including that which he encountered in the places he visited.

Before the introduction of coke, the best fuel for malt kilns was considered to be culm or Welsh coal from Swansea. It burnt without smoke and the sulphurous fumes from it bleached the malt to an acceptable degree of paleness. After coal-dried malt, straw-dried malt was said to be the least tainted, then wood- dried and then fern-dried. Fern was often used for cheapness because it could be cut on commons, but it gave the malt a disagreeable taste. Wood was commonly used although the taste it produced was said to be disagreeable if one was not used to it. However, it was widely used in some parts of the country such as the West Country. The wood which gave the best flavour was birch, followed by beech. It was essential to ensure that it was not green or wet in

ANDREW BORDE

order to reduce the amount of smoke. Harrison made the following comments on the use of kiln fuels in the sixteenth century:

> For the wood dried malt when it is brued, beside that the drinke is high of colour, it dooth hurt and annoie the head of him that is not used thereto, bicause of the smoake. Such also as use both indifferentlie doo barke, cleave, and drie their wood in an oven, thereby to remove all moisture that should procure the fume, and this malt is in the second place, and with the same likewise, that which is made with dried firze, broome, etc.: whereas if they also be occupied greene, they are in a maner so prejudicial to the corne as is the moist wood.

In some cases a smoky flavour became quite acceptable in certain beers. The Welsh ale Cwrw which was brewed from Saxon times right up until the nineteenth century was brewed with a smoky flavoured malt. This was said to render the beer 'glutinous, heady and soporific'. Smoked malt was also used in Norway and Germany and in the case of the later country a smoked malt is still produced today.

Smoke contamination of malt was not the only cause for concern. The whole process of malting requires great skill, and judging from the contents of a Bill passed in the reign of Edward VI (1547–1553) there were several aspects in which maltsters were showing shortcomings. The Bill was particularly concerned over short periods of germination which would give an undermodified malt. In this kind of malt, the chemical changes which transform the barley into malt are not completed. The bill states:

> That where divers and sundry persons taking upon them the art and mystery of maltmaking, and sundry other persons, tendering more their own private lucre, gain the profit, than the wholesale victualling of the King's Majesty, the nobility and other of His Grace's subjects, have of late by their unsatiable, covetous and greedy minds, accustomably and commonly made much malt impure and unseasonable, for that they have made the same in eight or nine days, when indeed the same can not be well and perfectly made unless it have the space of twenty-one days in the making thereof. . .to the great peril of the King's subjects. . .it is enacted that –
>
> No barley malt is to be made for sale unless the malt shall be in making and drying three weeks at least, except in June, July and August, and then it shall be in operation seventeen days at least, upon forfeiture of two shillings for every quarter. No person, except for the use of his own household, is to mix any malt not well made, or made of mow-burnt or spired barley, with other good malt for sale, upon forfeiture of two shillings for every quarter. No person to sell any malt not well trodden, rubbed and fanned, whereby there may be conveniently fanned out of one quarter, half a peck of dust or more, incurs a forfeiture of twenty pence per quarter.

WATER

The other big problem facing the brewer in his attempts to produce a beer of good quality was to find a reliable supply of water.

Harrison also had comment on the quality of water used for brewing and he wrote:

> In this trade also our Bruers observe verie diligentlie the nature of the water, which they dailie occupy, and soile through which it passeth, for all waters are not of like goodnesse, sith the fattest standing water is alwaies the best; for although the waters that run by chalke and cledgie soiles be good, and next unto the Thames water which is the most excellent, yet the water that standeth in either of these is the best for us that dwell in the countrie, as whereon the sunne lyeth longest, and fattest fish are bred. But of all other the fennie and morish is the worst, and the clerist spring water next unto it.

However, there were indications even in the reign of Henry VI (1422–1461) that the water of the Thames in London was not always as excellent as Harrison indicated. The Wardens of the Brewers Company were commanded not to take water from the Thames when it was disturbed, but to wait for low water and the turn of the tide. In the time of Elizabeth I (1558–1603) the river had begun to acquire a bad reputation. In 1592 the author of *Pierce Penilesse, his supplication to the Devill*, in reference to the London brewers, writes: 'Some are raised by corrupt water, as gnats, to which we may liken brewers, that, by retayling filthie Thames water, come in a few yeres to be worth fortie or fiftie thousand pounds.'

Outside London, brewers also used river water or water from ponds or streams. The temptation was to take it from the nearest source, but care was needed to avoid animal pollution and places where there was rotting vegetation. One writer points out the necessity to sieve out the small red worms when using pond water which had become low in summer. Well waters which were perhaps the purest of waters available were not widely used, partly because of the cost of raising the large amounts required and partly because they were often hard. The problems that they caused with the scaling of utensils and a belief that they had a similar effect on the body predisposed people against their use. Generally such hardness was temporary hardness, but supplies which were permanently hard because of their gypsum content were recognised as being of value for producing good beer. The monks of Burton-on-Trent did not know that their water contained gypsum but were aware that it produced particularly good beer.

The problems with water were not only confined to the British Isles, as it is recorded that the brewing of beer with foul water also occurred in Germany in late Mediaeval times.

HOPS

Hops were introduced into England from Flanders in 1400 and though by the seventeenth century had become the accepted flavouring agent of beer, they were unpopular for many years. They were regarded by many as an adulterant.

James Howell writing in a letter in 1634, describes the drinks in various countries which he has visited and of England he says:

> In this Island, the old drink was ale, noble ale, than which, as I heard a great foreign doctor affirm, there is no liquor that more increaseth the radical moisture, and preserves the natural heat, which are two pillars which support

> the life of man. But since beer has popped in amongst us, ale is sought to be much adulterated.

Andrew Borde certainly did not think much of the hop judging from his descriptions of ale and beer. Of beer he wrote:

> Bere is made of malte, of hoppes and water: it is a naturall drynke for a Dutche man. And nowe of late dayes it is moche used in Englande to the detryment of many Englysshe men; specyally it kylleth them the which be troubled with the colycke and the stone and the strangulion; for the drynke is a cold drynke; yet it doth make a man fat and doth inflate the bely, as it dothe appeare by the Dutche mens faces and belyes.

Of ale he wrote:

> Ale is made of malte and water; and they that do put any other thynge to ale than is rehersed, except yest, barme or godesgood, doth sofysticat theyr ale. Ale for an Englysshe man is a naturall drynke.

Henry VIII did not like hops either and in 1530 instructed his brewer at Eltham not to put hops or brimstone into beer brewed for royal consumption.

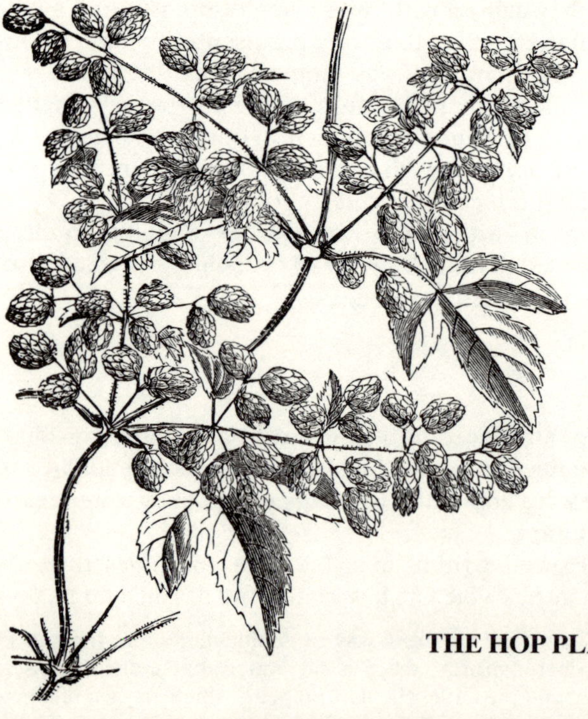

THE HOP PLANT

There were various prohibitions and protests over the use of hops. They were forbidden in Norwich in 1471 and in Shrewsbury in 1519, where they were described as a wicked and pernicious weed. In London, in 1484, the Brewers Company asked that no 'hops, herbs or other like thing be put into any ale or any liquor whereof ale shall be made but only liquor, malt and yeast'. Most hops until the later part of the sixteenth century were imported and one of the reasons that they were unpopular to start with was that they were frequently adulterated. In 1522 the Lord Mayor of London employed searchers in the port to find the unwholesome hops and destroy them. In 1554 hop growing in Britain was legalised by Act of Parliament indicating that hops were here to stay.

Hops were used in the south of England more than anywhere else as they were originally imported into Kent, which eventually became the main centre for their cultivation. Northern beers were more lightly hopped and more herbal beers were brewed. In 1630 John Taylor was surprised when he visited a Manchester man to find that he had nine different beers at his table including those flavoured with hyssop, wormwood, sage, rosemary, betony and scurvey grass as well as hops.

For many years not everyone was convinced of the goodness of hops and as late as the seventeenth century a petition was presented to Parliament by some citizens of London for the abolition of certain nuisances: 'Coal which gives a bad smell and hops, which destroy the flavour of the beer and are detrimental to the health.'

CHAPTER 2

ADULTERATION FROM 1700 TO 1900

In the eighteenth and early nineteenth centuries there was considerable adulteration of beer, the practice reaching a peak in the second decade of the nineteenth century. The situation deteriorated to a point where there was a serious health risk to beer drinkers. It could be fairly stated that most beer was not adulterated with any intention of grand fraud on the part of the brewer, but could be attributed to what one writer of the time described as a 'low standard of the art'. The implications of some of the brewing practices which we would find undesirable today were not fully appreciated with the knowledge available to the average brewer at the time. Very few brewers had any scientific training and were often unaware of any of the undesirable properties of substances which were sold to them by brewers' druggists and the like. Substances were offered for sale which for instance could cure their sour beer or could be used to replace hops, which sometimes more than doubled in price from one year to another. The large brewer was in the best position financially and could ride the storm when there were problems with the beer or if the raw materials soared in price; but the small brewer had to face these problems head on and use whatever means were available in order to overcome them. The small brewers would have been for the most part only too pleased to find some substance for sale which would help them overcome the problems they encountered, without going too deeply into investigating any of its other properties.

There was of course sometimes deliberate intention on the part of some brewers to deceive the customer , but one should not forget the role of the brewers' druggists who would be more aware of the unsuitability for human consumption of the substances which he was selling. The hard economic conditions around the times of wars with the French should also be appreciated, as they were not softened by the type of sophisticated economic control we have today.

LEGISLATION

It is perhaps surprising that beer was allowed to be adulterated in this period as various laws had been passed prohibiting the addition of certain substances.

In 1689 the use of molasses, coarse sugar, honey or any composition or extract of sugar was prohibited for use in brewing with the forfeiture of the beer and a fine of £100 as penalty for their use.

In 1710 the use of broom, wormwood or other bitter ingredients instead of hops was prohibited with a fine of £20 as penalty for their use. However, a retailer was allowed to add an infusion of wormwood or broom in order to make wormwood or broom ale.

In 1713 the use of sugar, honey, foreign grains, Guinea pepper, *essentia bine*, *Cocculus indicus*, or any unwholesome ingredients whatsoever was prohibited with a fine of £20 as penalty for their use.

In 1802 beer grounds, stale beer, sugar water, distiller's spent wash, sugar, molasses, vitriol, quassia, *Cocculus indicus*, grains of paradise, Guinea pepper, opium or any material or ingredient except malt and hops, was prohibited in brewing or in the preparation of any liquor added to beer. The penalty for use was a fine of £200 and forfeiture of the brewing equipment.

In 1816 any liquor, extract, calx or other material or preparation for darkening or colouring worts or beer was prohibited for brewing other than brown malt. Also prohibited was molasses, honey, liquorice, vitriol, quassia, *Cocculus indicus*, grains of paradise, Guinea pepper, opium or any substitute for malt or hops. The penalty for use was a fine of £200 and forfeiture of the beer and containers. Legislation against the suppliers, who were the bigger culprits, was also made. If any chemist, grocer or druggist was caught supplying these substances to brewers he was fined £500.

HOP SUBSTITUTES

One of the groups of adulterants which was most widely used was hop substitutes. The following table which was printed by order of the House of Commons in 1819 shows the fluctuation of the price of hops over the previous thirty years. This was the price per cwt of 'bag hops' which were the cheapest on the market. It explains, as already mentioned, why some brewers resorted to hop substitutes.

1789	Oct	£ 6.14s	to	£ 7. 2s
1790	Oct	£ 3.16s	to	£ 4. 4s
1791	Oct	£ 5. 5s	to	£ 5.12s
1792	Oct	£ 4.00s	to	£ 4. 4s

1793	Oct	£ 9. 9s	to	£10.00s
1794	Oct	£ 4. 4s		
1795	Oct	£ 5.00s	to	£ 5.12s
1796	Oct	£ 4.00s	to	£ 4. 4s
1797	Mar	£ 5.12s	to	£ 6.00s
1797	Oct	£ 4. 4s	to	£ 4.10s
1798	Oct	£ 9. 9s	to	£10.00s
1799	Nov	£14.14s		
1801	Jan	£16. 5s	to	£18.00s
1802	Jan	£ 5. 5s		
1803	Jul	£ 5.12s		
1804	Jul	£ 4.15s	to	£ 5.00s
1812	Nov	£13.13s	to	£15. 5s
1813	Jan	£13.13s	to	£15.10s
1814	Dec	£ 9. 4s	to	£ 9. 9s
1815	Feb	£ 8.16s	to	£ 9. 9s
1816	Jul	£ 6.10s	to	£16.17s
1816	Oct	£14.14s		
1817	Jan	£14. 5s		
1818	Jan	£31.00s		

In 1796 the use of hop substitutes had reached such proportions that the hop planters put the following advertisement in the *Monthly Magazine* for August:

> Kent. At a meeting of the Hop-Planters lately held at Maidstone, Sir William I. Twysden in the chair; it was resolved that a reward of £200, should be given to any person who shall detect any brewer in the use of unlawful ingredients as substitutes for hops: a committee of forty gentlemen was appointed to carry this resolution into effect. Beer was formerly a valuable beverage, composed of water, malt and hops; but there is now too much reason to believe, that much of it is a poisonous preparation, consisting of Molasses, Spanish-juice, Elderberry-juice, Ginger, Quassia-wood, or its tincture, Cocculus Indicus (which poisons fish, and stupifies man), with but little malt and hops.

The steep rise in price of hops in the previous year, as seen from the above table, had doubtless caused an upsurge in the use of substitutes. The growers' anger was aimed at the brewers for using the substitutes, but the real culprit was one man, who had monopolised the trade in hops and pushed up the price. However, such was the public concern over the matter that he was brought before the Court of King's Bench where he was found guilty of 'monopolising hops and anticipating the markets'. This was not an unknown practice and had occurred before in 1551 when four regrators or forestallers, as they were called, had accumulated large amounts of hops in a time of scarcity in order to push up the price. On that occasion they were ordered to sell their stock at a reasonable price. The price of hops was again pushed up at a later date in 1883 but by a more subtle means. A firm of hop merchants leaked out the fact that they had calculated that there was a deficit of 245,000 cwt in the amount of hops required by the industry for that year. The unfortunate effect was not to run up the price of hops but to provide ammunition for the temperance movement, who stated that the said deficit was made up by use of 245,000 cwt of chemicals.

One of the reasons why brewers could get away with using hop substitutes in their beer was that most of the beer produced, particularly in London, was porter. This was a fairly robust beer in which any of the subtleties in flavour of fine hops were lost and generally the oldest and coarsest hops were used to brew it. The finest of hops were used in ales. Substitutes which merely gave the high amount of bitterness required in porter were regarded as being satisfactory. However, the use of substitutes may have been a miscalculation on the part of brewers, as with other factors it probably lead to the eventual demise in popularity of the drink.

OTHER ADULTERANTS

Hop substitutes were not the only substances added to porter. A typical recipe is taken from the *Complete Family Brewer* by Thomas Threale published in 1802:

1 quarter of malt
8 pounds of hops
9 pounds of treacle
8 pounds of liquorice root
8 pounds of colour
2 ounces of Spanish liquorice
¼ ounce of Creolus Indian Berry
2 drachms of salt of tartar
¼ ounce of powdered alum and copperas,
    half and half, mixed
3 ounces of ginger
4 ounces of slacked lime
1 ounce of linseed
2 drachms of cinnamon

This made five barrels of porter. Details of the use of these ingredients is given later but at this point the list is used to illustrate the range of ingredients which could be used to brew beer at the time. Admittedly this quotation was taken from a book intended for the 'private' brewer as opposed to common brewer, so the author could recommend a whole range of ingredients which were illegal other than for private use. However, many of these adulterants along with others were also included in the recipes recommended for the common brewer by other writers of the time.

A point worth mentioning in relation to this is that Threale and other such brewers who wrote technical brewing books must have had a genuine belief that these ingredients were not harmful, as there was no advantage in using them for domestic brewing unless the intention was to improved the product. Financial savings made by using these ingredients on a domestic scale would have been minimal and their use must have been through well-intended ignorance on the part of the brewer. Some people were fortunately a little more enlightened. Dr Buchan in his *Domestic Medicine* written around the same time, makes the following attack on brewers and urged families to brew their own beer in order to avoid the adulterants used in commercially brewed beers.

> All families who can, ought to prepare their own liquors. Since preparing and vending of liquors became one of the most general branches of the business, every method has been tried to adulterate them. The great object both to the makers and vendors of liquors, is, to render it intoxicating, and give it the appearance of age. But it is well known that this may be done by other ingredients than those which ought to be used to make it strong. It would be imprudent even to name the ingredients which are made use of to render liquors heady. Suffice it to say, that the practice is very common, and that all the ingredients used for this purpose are of a narcotic, stupefactive quality; but as all opiates are deleterious, it is easy to see what must be the consequence of their general use. Though they do not kill suddenly, yet they hurt the nerves, relax and weaken the stomach, and spoil the digestion. Were fermented liquors faithfully prepared, kept to a proper age, and used in moderation, they would prove great cordials to mankind; but, while they are ill prepared, and adulterated, and taken to excess, they must have many pernicious effects.

He would not name the substances used for fear of perpetuating their use, but they were well documented elsewhere at the time.

PUBLIC CONCERN

The first display of any public concern was just prior to 1818, when a petition complaining about the inferior quality and high price of beer was presented to Parliament by a group of people of London and the vicinity. The complaint was investigated for the period covering the six years preceding the date of the petition, by a Select Committee who published their Report in June 1818. They concluded that the eleven largest brewers in London were using legal ingredients during the period examined, with the exception of one instance at Messrs Thomas Meux's house. In 1812 he had been convicted of using salt of tartar to correct an acidity in a young beer. It was said to have been carried out in all innocence by the chemist employed, who believed it to possess no pernicious qualities. They further concluded that it was the smaller brewers and publicans, both in the Metropolis and in the country, who were using adulterants such as hop substitutes. However, the production of porter was in the control of the large London brewers and it was this type of beer which technical writers of the time indicated to be the most commonly adulterated. There may have been some political pressure to exonerate the large London brewers as many had political connections. Such was their influence as a group they were commonly known as the 'beerage'. The Report does seem to labour the point that the large London brewers were not involved in the practices and even goes as far as to sympathise with them saying that their reputations must have been sullied by the actions of the other brewers. However, there is no doubt that the smaller brewers were good customers of the brewers' druggists, as Fredrick Accum writing two years later confirms the fact.

The information is given in his *Treatise on the Adulteration of Food and Culinary Poisons* published in 1820 in which he did his best to spread public awareness of the widespread practice of adulterating food, including beer. The cover of the book itself was enough to alarm some people and the description given by one reviewer is as follows:

> Mr Accum seems determined that even the outside of his book shall awaken all our fears. The cover of our copy bears a death's head emblazoned upon a pall, and, underneath, the motto "there is death in the pot". The pall is supported by the point of a dart. Four other darts support the four corners of the device. Twelve serpents, with forked tongues and tails entwined, form a terrific wreath around; while the middle is occupied with a large cobweb, delineated with much attention to detail, in the centre of which is a spider, full as large as a moderate sized hazel nut, and so frightful that more than one young lady of our acquaintance would think it necessary to scream at the sight of it, holds in its envenomed fangs an ill fated fly, which is sinking under the loss of blood, and buzzing in the agonies of death.

On reading the book Accum was found to be no less forceful with his words, and the contents so alarmed another reviewer that he wrote:

> Since we read his book, our appetite has visibly decreased. At the Celtic Club, yesterday, we dined almost entirely on roast beef . . . Our diet is at present quite similar to that of Parnel's hermit,
>
> "Our food the fruits; our drink the crystal well;"
>
> though we trust a few days will recover us from our panic, and enable us to resume our former habits of life.

Accum, unlike Buchan, details the offending ingredients, particularly deploring the use of narcotic substances used to make beer more intoxicating. He tells of the rise of the brewers' druggist. The substances which he sold were not all harmful but many were poisonous as seen from the following quotation:

> To increase the intoxicating quality of beer, the deleterious vegetable substance, called cocculus indicus, and the extract of this poisonous berry, technically called black extract, or by some, hard multum, are employed. Opium, tobacco, nux vomica, and extracts of poppies have also been used.This fraud constitutes by far the most censurable offence committed by the unprincipled brewers. And it is a lamentable reflection to behold so great a number of brewers prosecuted and convicted of this crime. Nor is it less deplorable to find the names of druggists, eminent in trade, implicated in the fraud, by selling the unlawful ingredients to brewers for fraudulent purposes.

He then adds the following comments:

> That a minute portion of an unwholesome ingredient, daily taken in beer, cannot fail to be productive of mischief, admits of no doubt: and there is reason to believe that a small quantity of a narcotic substance (and cocculus indicus is a powerful narcotic), daily taken into the stomach, together with an intoxicating liquor, is highly more efficacious than it would be without the liquor. The effect may be gradual; and a strong constitution, especially if it be assisted with constant and hard labour, may counteract the destructive consequences perhaps for many years. But it never fails to shew its baneful effects at last.

The following is his list of the adulterants seized from brewers at the time:

> Multum, 84 lbs.; cocculus indicus, 12 lbs.; colouring 4 galls.; honey, about 180 lbs.; hartshorn shavings, 14 lbs.; Spanish juice, 46 lbs.; orange powder, 17 lbs.; ginger, 56 lbs.; grains of paradise, 44 lbs.; quassia, 10 lbs.; liquorice, 64 lbs.; carraway seeds, 40 lbs. ... capsicum, 88 lbs.; copperas, 310 lbs.; colouring and drugs, 84 lbs.; mixed drugs, 240 lbs.; coriander seed, 2 lbs.; beer colouring, 24 galls.

THE CONTINUED USE OF ADULTERANTS

Most of the adulterants mentioned by Accum were recommended ingredients in a brewing textbook called *A Practical Treatise on Brewing* written by Morrice and published in 1802. Editions were still being published at the time that Accum was writing and even as late as 1834. This does seem to have been a strange situation in retrospect, but perhaps the same will be said in the future about the situation in the late twentieth century.

The use of adulterants continued despite the publicity between 1818 and 1820 and is confirmed by Bonnington Mowbray, writing in 1824, in his book, *The Private Brewery*. He stated:

> It has seldom been my fortune, in a great number of years, to taste unadulterated purchased ale, whether brewed in the Metropolis, or in the brewing districts of the country. In these ales, at present the chief articles of adulteration, which my well-practised palate can discover, are seeds, sugar and salt, perhaps bay salt, Cocculus Indicus (Indian Berry), a most intoxicating and deleterious drug, the flavour of which I well know, was formerly much in use; and if that be not the case at present, it is difficult to conceive why such large quantities of that drug should be annually imported, since its use in medicine is almost unknown. The brewers, however, find a sufficient apology in the pressure of taxation, and the vitiated taste of the public.

Their use was still continuing five years later, as David Booth in his *Art of Brewing* gives quite a detailed account of the illegal substances used in beer. He concludes it with the following two paragraphs:

> It will be said that every article which we have here stigmatized is medicinal, and appears in the pharmacopoeias; but we also know that there is no substance, however deleterious or disgusting, which has not, at one time or other, found a place in the Materia Medica. Besides, the parallel is imperfect. In medicine the poison is prescribed in measured doses, (less or more, according to the prudence or rashness of the physician) and only in such diseases as are deemed incurable; whereas the brewer, or his drayman, administers the drugs without discrimination, ignorant and careless of the age, sex, or constitution of his patient.
>
> In the tone of reprehension, which we have felt it our duty to assume on this subject, we trust that we shall not be accused of personality. Let it be remembered that we address ourselves to the most worthless of the trade, to such as disgrace the name of brewer, by sporting with the lives of their fellow creatures, for the sake of gain. If there be any honest man so weak as to suppose that we mean to throw suspicion on the brewery in general, we wish him to be undeceived. We are willing to believe that the number of reckless beings who use deleterious ingredients are few; but that there exist those few, is too well ascertained, from the seizures and convictions that have been so often made, and are still making, by the Excise. Our denunciations are directed solely against the guilty; and sorry should we be, if they could possibly be conceived to allude to any respectable House, or to any honourable man.

Rather strangely in the second part of the book published later in the same year, he gave details of brewing Burton ale and listed as materials to be used in the beer, some of the illegal ingredients mentioned in the first part. This smacked of the double standards which had existed for decades, but Booth had not reckoned with the raised standards of the Burton brewers. Their light, more delicately flavoured beers could not withstand the use of the adulterants used in porter. The absence of adulterants, the superiority of their brewing liquor and technical advances in brewing such as attemperation, had caused the reputation of their ales to grow. They were quite intent on maintaining this and decided to sue the publishers for libel, affirming that they did not use any of the deleterious and disgusting ingredients mentioned, but only malt, hops and water, though they did concede to the use of flour and salt, and jalup.

The details of the court case were published in the last part of the book which was published after it and shows that the Burton brewers' reputation was upheld. However, it was clear that Mr Booth was not entirely convinced of their innocence. The honesty of the Burton brewers apart, it does show that there was some movement in the brewing trade towards halting the use of illegal ingredients.

Booth does seem to have been a bit unfortunate, as he had made a most apologetic attack compared to the more vitriolic attack made by Brande in his *Town and Country Brewery Book* published at about the same time in 1830. He was, however, careful enough not to identify or give information about any group of brewers in particular but generalised his comments. The following are some extracts from what he wrote:

> British beer . . . has always claimed the superiority over that produced by any other country, when divested of the sophisticating ingredients with which, at the present day, it unfortunately is too frequently combined. But it will scarcely admit of a belief that there are few articles of consumption, in which, hitherto, the good people of old England have been more grossly imposed upon, or in which they have suffered more by the fastidiousness of pretended taste, than in the instance of malt liquors. The poisonous and nauseous drugs with which these otherwise salutary and nutritious drinks, particularly, as Fletcher calls one of them,
>
> "Strong, lusty, London beer,"
>
> are saturated, or, to use the technical phrase of the trade, doctored, would almost argue that John Bull is poison proof: for in most brewers' ales and porters what do we find? Why, (with horror we divulge it,) opium, henbane, cocculus indicus, grains of paradise and Bohemian rosemary; which latter is said to produce a quick and raving intoxication; —these are made to supply, from their heady and narcotic properties, the place of alcohol, or the spirit that ought solely to be produced by the malt. Aloes, gentian, quassia, sweet scented flag (orrice root), wormwood, horehound and bitter orange peel are used as substitutes for hops.
>
> Liquorice-root, treacle, and mucilage of flax seed, stand for unattenuated malt sugar. Capsicum, cassia buds, and ginger afford to the exhausted and badly managed drink, all the pungency of carbonic acid.
>
> Burnt flour, sugar or treacle embody a peculiar taste fancied by many; and

isinglass, cods'-sounds, and, in obstinate cases oil of vitriol (sulphuric acid) render them transparent.

The brewer, independent of the above long and black catalogue of secret poisons, (we do not allude to them all as coming under this denomination) has occasionally to supply himself with potash, lime, salt, and a number of other materials, which, indeed, are of no other harm than as illicitly ejecting those of a more wholesome and valuable nature; to the great loss and detriment of the consumer. An Act of Parliament was certainly passed to prevent the use of these noxious drugs, under a penalty which is by no means adequate to the flagrancy of the crime; but, if the legislature would carry the penalty to the greatest possible extent,– nay, make it even felony, without benefit of the clergy, to detect any one engaged in brewing, in the use or the possession of poisonous ingredients; no one, who valued the interests of society, individually and collectively, would murmur at the severity of the decree.

The publican comes in next, and justly so, for his share of the obloquy which he would not prevent. His heading, his flavour and bulk are all poisons and impositions on the public, which the wisdom of the legislature has at length seen through; and at the expence of the enormous reduction of the revenue, has struck at the root of this long-lasting evil, which has consigned thousands of his majesty's poorest subjects, long before their time, to their last account.

THE DUKE OF WELLINGTON'S BEERHOUSE ACT

In the year 1830 an important piece of legislation was passed by Parliament which came to be known as the Duke of Wellington's Beerhouse Act. It made two important changes:

Firstly, it abolished the duty on beer which was paid by the common brewer but not the private brewer, putting the burden of taxation on malt which was used by both. The private brewers tended to be the wealthier classes who had their own private brewhouses and could brew duty-free beer, whilst the poorer classes bought their beer from the common brewer, who passed on the cost of duty in the price of his beer. It was believed that once some of the burden of taxation was taken from the common brewer, he would stop trying to make financial savings by using adulterants as substitutes for wholesome materials.

The second change it made was to open up the beer trade, which in effect meant that anyone could retail beer (not spirits) on the payment of two guineas per year. The main idea was to curb spirit drinking which had become a social problem because of the ease with which it caused high levels of drunkenness. Beer, because the alcohol is more dilute, does not so easily cause drunkenness and was regarded as being relatively wholesome. This is well illustated in the famous engravings of Hogarth in which Gin Lane is shown as a scene of debauchery, death and poverty whereas Beer Street is a scene of prosperity, happiness and life. It was also believed that competition would be increased between brewers and less adulterants would be used in an attempt to produce a better quality product.

Brande makes the following comment on the Act:

> So then it has come to this at last, the beer trade is at length thrown open, to secure a free, and we trust, an honourable and praiseworthy competition, among brewers, publicans, and venders of beer, of every class: it is by such means, only, that the long wanted antidote to the long existing gangrene can be effectually applied, and the evil eradicated; for fortunately, there no longer exists a doubt in the public mind of the pernicious practices so remorselessly carried on in the beer trade. These ought no longer to be a secret, that their detection may be the more readily exposed, and when these secrets are known, we need only repeat the words of the Chancellor of the Exchequer: "Let any man who (hereafter) drugs his beer, beware of the competition which will soon rob him of all his customers," and, we may add, consign him to merited punishment.

Alas, it did not quite work out as had been expected. The beer houses were under fierce competition from each other because so many opened up and in turn they were all in competition with the houses which had a full licence and could sell spirits. Spirits brought in a good profit so the beer houses were at a serious disadvantage when trading in competition with houses with a full licence. This in turn led to malpractices by some beer houses in order to survive. Measures were taken in 1834 and 1840 to improve the situation by introducing greater control in the setting up of beer houses, with the introduction of further Beerhouse Acts. Eventually in 1869 the issue of all retail licences was put in the hands of the licensing justices.

INCREASED CONTROL

By 1840 the problem of adulteration was more under control, and during the remaining years of the nineteenth century new measures were periodically introduced which, apart from the practice of dilution, completely eradicated the problem.

The situation in 1840 was that although the legislation existed to prosecute the adulterators of beer it was often difficult to prove adulteration had taken place. Dishonest publicans and brewers well knew this and providing that they exercised a little care in the use of their adulterants, they avoided detection. The first move to improve the chances of prosecuting the adulterators was the establishment of the Inland Revenue Laboratory in 1843. The First Annual Report of the Commissioners of Inland Revenue in 1857 gave an account of the first thirteen years work of the Laboratory.

It was reported that of 215 samples of beer taken between 1844 and 1856, 142 showed signs of adulteration. Of sixty-eight samples of hops taken, fifty-six contained adulterants. The report stated that:

> All the above samples were selected, on account of their apparently illicit character, from the premises of licensed Brewers or Sellers of Beer.
>
> The substances usually used in the adulteration of Beer, and which have been found in the possession of licensed Brewers are, Cocculus Indicus, Tobacco, Grains of Paradise, Chillies, Caramel, Coriander seeds, Quassia, Chiretta, Gentian root, Saltpetre, Carbonate of Soda, Sulphuric Acid, Common Salt, and an extract containing Aloes, Liquorice, Caramel and Sulphate of Iron.
>
> The substances most generally found in Hops which have been used by Brewers are Cocculus Indicus, Grains of Paradise, Quassia, Chiretta, Gentian, Chamomile flowers, Coriander seeds, and, in one instance, exhausted Tobacco.

In 1860 the Government introduced legislation known as the Adulteration of Food Act, which was the first serious attempt to protect the public from exploitation by dishonest traders. Its powers were limited to controlling the sale of adulterated food which was aimed at 'defrauding or injuring the health of Her Majestie's Subjects', but at least it was a beginning. More effective control was another twelve years in coming with the passing of the Adulteration of Food Act 1872, which was fully titled An Act to Amend the Law for the Prevention of Adulteration of Food and Drink and of Drugs. The main amendment was that as well as being illegal to sell adulterated food it became illegal for anyone to admix adulterating ingredients or to cause others to do so, during the manufacture or preparation of food. The emphasis on stopping the people who were producing the adulterated foodstuffs was made clear, as the penalties for it were far more severe than the penalties for selling the adulterated products. The Act also made it compulsory for Local Authorities to appoint public analysts whereas before it had only been a recommendation, and it required them to instigate proceedings against anyone suspected of producing or selling adulterated food, whereas under the 1860 Act it was necessary for the purchaser to instigate proceedings and to pay the fee for analysis.

As far as intoxicating drink was concerned provision was made under the Licensing Act of 1872 for the analysis of beer suspected of being adulterated and for the prosecution of offenders. This Act was passed at the same time as the Adulteration of Food Act so that the two worked together. Specific substances were named as being prohibited for use in beer and included *Cocculus indicus*, sodium chloride, copperas, opium, Indian hemp, strychnine, tobacco, darnel seed, logwood extract, salts of zinc, lead or alum and any other deleterious ingredients. Samples had to be secured by a police officer or an officer of the Inland Revenue and analysed by someone appointed by the Commissioners of the Inland Revenue. It appeared that the public analyst was not to be very involved in detecting the adulteration of beer as the Inland Revenue had their own Laboratory at Somerset House and had the experience in analysing beer for adulterants since 1843. This part of the Act was repealed in July 1874, perhaps because it was felt that there was adequate provision to deal with cases of beer adulteration under the Adulteration of Food Act 1872. The new 1874 Licensing Act did, however, provide for any convictions for adulterating beer to be recorded on a licensed person's licence.

Despite the introduction of legislation, adulteration was still continuing and in the Inland Revenue Report for 1871–72, it as stated that nineteen samples of beer were collected during the period as they were suspected of being adulterated. Five samples proved to be genuine and fourteen were found to be adulterated. Ten were adulterated with substances which were injurious to health or likely to promote intoxication and of those ten, all were confirmed to have been taken from small brewers in the Midlands. The figures of course only show those cases which were detected, but at least they indicated that the use of harmful ingredients was still continuing.

In a report of the Local Government Board on the working of the Sale of Food and Drugs Act, it was reported that in 1880, 465 samples of beer were analysed by public analysts, of which nineteen were found to be adulterated amounting to less than 4% of the total. The 127 samples taken in the Metropolis were found to be free of adulterants. Those which were found to be adulterated came from outside the Metropolis and were found to contain only salt. In 1885 the level of detection had fallen to 2.8% and the adulterants were again all salt.

SALT

There had been public concern for some time that brewers were adding large amounts of salt to beer in order to make the drinker thirsty, so that more beer could be sold. The credibility of this story was obviously very high in the minds of those who wished to believe it. The *Illustrated London News* for example, in 1850, carried a short piece on the practice of 'Bobbing beer', with an illustration. This showed two men supposedly adding blocks of salt to some small vats of beer. A little knowledge of

BOBBING THE BEER

brewing would show that such a large amount of salt would have made the beer undrinkable. What the artist had drawn could have been blocks of sugar which were mistaken for blocks of salt. At the time of the introduction of the Adulteration of Food Act in 1972, the *Daily Telegraph* decided to bring to the attention of the public the practice, and in a leading article asserted that 'salt is extensively used, but mainly, perhaps, for the purpose of inducing thirst, a most rascally practice, and one which should be visited with the severest of penalties'.

The Excise were mainly concerned with seeing that the revenue was not affected by the use of adulterants and as salt offered no threat to the revenue, because it was not used as substitute for malt or hops, it was decided that the Public Analysts should be responsible for its detection in beer and for any subsequent prosecutions. The Excise had only previously prosecuted for adulteration with salt if levels were above 50 grains per gallon, but had been fairly flexible realising that many water supplies especially in coastal districts contained levels of salt well above this mark. There was nothing at all the average brewer in those days could do about the levels in water supplies, but generally it little affected the quality of his beer. The other problem was that salt levels could build up in a beer during brewing without the brewer being particularly aware of the fact. In particular, levels could increase through concentration during boiling. These could be quite significant where there were high rates of evaporation and long boiling times were used. An example of this was illustrated by Dr Alfred Hill, in 1872, when he analysed a brewing liquor and found 22.5 grains per gallon and 52 grains per gallon in the beer brewed from it, although no salt had been added. It was also common practice to dress fermentations with flour and salt. This had the physical effect of releasing gas from the beer which carried the yeast up to the surface of the brew so that a good head was formed for skimming. Casual measurements of the amounts of salt used for dressing could increase levels more than desired.

Some public analysts were a little inflexible in their approach to the matter and having decided on the level that they would allow they prosecuted any brewer who had levels in his beer which were above it. In the *Lancet*, in 1882, appeared a report that the public analyst for St George's-in-the-East, London, had 'met with twenty four samples of beer highly charged with salt', which he denounced as 'being something worse than mere adulteration'. In a letter to the *Morning Advertiser* Dr H. C. Bartlett took up the case. He himself had taken samples of beer in the district four years earlier for analysis, on behalf of a philanthropic person concerned with the welfare of the poorer inhabitants of the area, and found no adulteration. On investigation he found that the public analyst's twenty-four samples were in fact two and that the maximum content of salt was 54 grains per gallon. The other twenty-two samples were other foodstuffs such as mustard and cocoa. He felt that since salt cannot be tasted in water below the level of 75 grains per gallon it was unlikely to induce a thirst in beer drinkers.

Another public analyst who tried to persecute the brewers was Mr Morgan the Public Analyst for Swansea. He caused much bitterness in the area by prosecuting brewers who had more than 50 grains per gallon of salt in their beer, labelling them as adulterators. In 1887 Messrs Giles & Harrap of Merthyr Tydfil were prosecuted and decided to clear their good name by contesting the action in court. They won their case and it was accepted that the high level in their beer was not a result of adulteration. The case attracted some national publicity and resulted in public analysts taking a more realistic attitude over salt levels in beer.

THE INDUSTRY UNDER ATTACK

Accusations of adulteration were often used by people or organisations in the later half of the nineteenth century as a weapon to attack the brewing industry for one reason or another. One of the great enemies of the industry was the Temperance Movement who, though they were fighting the evils which were brought about by drunkenness and alcoholism, were only too willing to point out how the brewers were cheating their customers and increasing the intoxicating effects of their beer by adding adulterants.

An example of such an attack was the one made by the Rev. Canon Bardsley in a report presented to the Convocation of York in 1872. He quoted a correspondent who wrote that on one occasion he heard an eminent brewer state 'that he carried his granary in his waistcoat pocket', implying that he had some potent drug which could be used to serve him in lieu of malt and possibly hops. Another example which illustrates that brewers were often accused of adding intoxicating drugs to make them more saleable, comes from the report of the Chaplin of Leeds Borough Gaol for 1871. He asserted that:

> Within the boundary of your worships' jurisdiction are to be found retail houses from which a customer, having partaken of one pint of malt liquor, or fourpenny worth of spirits may be sent home utterly incapable of taking care of himself.

Another story which was ultimately seized upon by the British temperance movement originated as an attack on the British brewing industry by the French. The surprising source of the attack was M. Payen, the noted French chemist who had discovered cellulose and the decolourising power of charcoal in sugar refining. He stated in a lecture at the Conservatoir des Arts et Metiers, that strychnine was prepared in large quantities in Paris, for exportation to England to be used in English ale, in order to give it bitterness. The story was reported in *Le Constitutional* and other French papers from where it was picked up by English newspapers. It caused quite a stir and was investigated by the Analytical and Sanitary Commission of the *Lancet*. Forty samples of beer were analysed and the Burton brewers threw open their stores for examination. The story, of course, proved to be quite unfounded and it was pointed out that the bitterness of strychnine and hops are quite different. In addition the degree of bitterness required in beer would have meant that it would have to have been added at twice the lethal dose. The French were not strangers to such tactics, which they saw as acts of patriotism rather than malicious slander.

Another example of such action was when the French decided to attack the Germans over the use of salicylic acid as a preservative. Somewhat deservedly, the attack backfired on them causing some harm to the French brewing industry. There had been some evidence that the Germans had been adding salicylic acid as a preservative to the beer which they exported to France. In 1885 the Prefect of Police in Paris ordered the Commissaries of Police on duty at railway stations to sample all beers arriving from abroad and the provinces. The samples were sent to the Municipal Laboratory for analysis. Within a week seven French beers and one German beer had been discovered to contain the preservative. There followed some controversial action over the fate of the beer discovered and the punishment of the brewers concerned. This attracted considerable attention from the press and there was much public debate on the adulteration of beer, which did the French brewers little good. Fortunately for British brewers, salicylic acid was little favoured by them and exports to France were not involved.

As already mentioned, some public analysts were a little over zealous with brewers, but one who was 'anxious to magnify his office' went to the extent of making totally unfounded accusations against the brewing industry. He stated that *Cocculus indicus* which was used at the beginning of the nineteenth century was still in use. In 1887 this lead to an investigation of the fate of imports. It was found that a small amount was available in most towns as it was a favourite drug used by poachers to stupefy fish. It was also the active ingredient of certain ointments for the destruction of parasites in cattle, but by far the majority of the drug was shipped out to other countries, as in those days Britain was carrier to the world. Germany and other European countries were the main destinations. No trace of use of the drug by a brewer, it was said, had been recorded for forty years.

Accusations were also made on a local level to protect certain interests. In Devon the cider makers and the brewers had always been in competition. The cider makers had suffered a setback in the eighteenth century when Dr Baker had proved that cider was the cause of an epidemic of Devon Colic, more scientifically known as lead poisoning. Unfortunately, the brewers' reputation for adulteration in the early nineteenth century did little to maintain any advantage that the problems of the cider makers had given them. The cider makers made sure that the public did not forget the brewers' malpractices. Without exaggeration, one of the best ways of doing this in Devon is to incorporate the point into local folk legend and it will be repeated over and over again for centuries, even on local radio. One of the best accounts of the legend was given by the well-known Devonshire historian, the Rev. Baring-Gould, in 1899.

> There are, as may have been noticed, three cold nights in May – not always, but often. At Crediton, and throughout the apple growing districts in North Devon, these are called 'Francémass' or 'St Frankin's days'; they are the 19th, 20th and 21st May. When a frost comes then it injures the apple blossom. The story relative to this frost varies slightly. According to one version there was an Exeter brewer, of the name of Frankin, who found that cider ran his ale so hard that he vowed his soul to the devil on the condition that he would send three frosty nights in May to annually cut off the apple blossom. The other version of the story is that the brewers in North Devon entered into a compact with the Evil One, and promised to put deleterious matter into their ale on condition that the devil should help them by killing off the blossom of the apple trees. Accordingly, whenever these May frosts come we know that his majesty is fulfilling his part of the contract because the brewers have fulfilled theirs by adulterating their beer. St Frankin, according to this version, is an euphemism for Satan.

Local brewers had to take such accounts seriously. When the Uffcombe Steam Brewery advertised in the *North Devon Journal* in the 1860s they emphasised to their potential customers that their beers were free from adulterants.

ADULTERATION DEFEATED

In 1875 the Sale of Food and Drugs Act was passed which consolidated previous legislation and specifically made it an offence to sell food which had been adulterated.

Gradually the adulteration of beer had been brought under control but there was concern amongst many people for the quality of beer when the Customs and Inland Revenue Act of 1880 was passed. It was basically an act to reform the method of taxing beer, but the opportunity was taken to allow brewers to use whatever materials they wished for brewing beer. The decision was based on grounds of political economy rather than any consideration for the consumer. Attention was focused on this aspect of the Act and using Gladstone's terminology it became commonly known as the Free Mash Tun Act.

However, picking up public concern, the Inland Revenue Commissioners were quick to dispel fears that beer would be adulterated and issued the following statement in their annual report for 1881:

> So far as we are aware, no attempts have been made to use materials in brewing at all detrimental to public health; and the presence of Revenue Officers in breweries affords fresh security to the public – if indeed any such were needed – against all such practices.

In 1885, due to increasing pressure, necessary legislation was introduced under Section 8 of the Customs and Inland Revenue Act in order to prevent abuse of the Free Mash Tun Act. It prohibited:

> . . . the adulteration of beer by brewers and the addition thereto of any matter or thing (except finings for the purpose of clarification, or other matter or thing sanctioned by the Board of Inland Revenue) before delivery for consumption.

The section also prohibited the dilution of beer by dealers or retailers.

In the year of introduction 314 prosecutions under this section were made and heard by the Metropolitan Police Magistrates. The defendants with one exception were convicted and the fines imposed totalled £7,637. However, in not one case was deleterious matter discovered and the adulteration was confined to sugar and water. The defendants were all publicans who had watered their beer and in some cases had tried to cover their tracks by adding sugar to compensate for the drop in gravity.

There are no records after this time of brewers or licensees being prosecuted for the addition of deleterious substances to beer. With the exception of dilution, the adulteration of beer seems to have been defeated. A parliamentary select committee investigating the use of brewing materials in 1896, following the attempted introduction of a Pure Beer Bill, confirmed this view.

CHAPTER 3

DILUTION AND WASTE BEER

DILUTION

The addition of water to beer is the simplest way in which it can be adulterated, making it the most commonly used adulterant in beer. It does no harm at all to the consumer except that it cheats him of what he has paid for. Adulterators often faced the problem of making their customers ill or even poisoning them, but this problem does not occur with water. It certainly has its attractions as an adulterant, so not surprisingly it is the only one which has survived in use to the present day. However, dilution has been confined to the retail trade. These businesses are usually under the control of one person and lend themselves more to clandestine activities than do breweries. Breweries usually have several employees who would have to be involved, making secrecy difficult. They are also regularly inspected by H. M. Customs & Excise with the intention of discovering such activities if they are occurring.

ALE CONNERS

In Mediaeval times it was the duty of the ale conner or ale taster to discover whether a brewer had produced beer which was too weak. The brewer, who was usually the retailer as well in those times, did not necessarily dilute his beer after he had brewed it, but diluted the wort taken from the mash tun. Brewers often must have diluted the wort as much they could without making the beer unacceptable, but sometimes they overdid it. The ale conner was the person to decide when this had happened. Dilution can be measured only when a standard has been established to which reference can be made. The method used has been described many times in brewing literature, but seems so improbable that no one is sure whether it is fact or fable. The ale conner is reputed to have poured the beer to be tested onto a wooden bench, then donning a pair of leather breeches he sat in the ale for half an hour. The sugar in the beer caused him to stick to the bench when he tried to rise and the amount of stickiness was used as a guide to the strength of the beer.

THE INLAND REVENUE

The ale conner was acting purely in the interests of the consumer but with the introduction of duty, the Inland Revenue took an interest in the dilution of beer. Initially, beer was classified as small or strong for duty purposes and the only risk of loss of duty was if small beer was passed off as strong beer, or more commonly if strong beer was diluted with small beer. When table beer was introduced as another classification between strong beer and small beer in 1782 it gave greater opportunity for dilution.

Suitable legislative precautions were taken to prevent the use of table beer as an adulterant of strong beer. Every cask containing table beer had to be marked with a large Roman letter 'T', which had to be at least four inches high. The beer also had to be kept in a separate storehouse or place, away from other beer. Failure to do this could result in a £50 fine. It was also forbidden to store it in a vessel larger than three barrels capacity. This was presumably to prevent it from being easily mixed with strong beer. Contravention of this regulation incurred a fine of £100.

The main problem that eventually arose was that table beer was being diluted or being substituted with small beer. In 1802 it was decided to merge the two classes and have only one class to be known as table beer. The duty on this beer was 2s. per barrel, whereas the duty on strong beer had been 3s. and on small beer had been 1/4d. However, the practice of dilution continued and in the 1819 session of Parliament an act was passed which prohibited it and stated:

> If any common brewer shall mix, or suffer to be mixed, any strong beer, or strong worts with table beer or table worts, or with water in any guile or fermenting tun after the declaration of the quantity of such guile shall have been made; or if he shall at any time mix, or suffer to be mixed, strong beer or strong worts with table beer worts or with water, in any vat, cask, tub, measures or utensil, not being an entered guile or fermenting tun, he shall forfeit £200.

The same prohibition applied to retailers but the fine imposed on them was only £50.

In the 1820s another type of fraud was discovered in Edinburgh which involved the dilution of beer. It was aimed at defrauding the Inland Revenue rather than the customer. Some brewers of small beer were brewing it a little stronger than normal and just before it was sent out, water was added to bring it to normal strength. In those times duty was paid on the volume collected for a particular type of beer and the specific gravity was not taken into account as it could not be measured, so the smaller the amount collected, the lower the duty bill. By diluting the beer the brewers were in effect obtaining duty-free beer from which they gained extra profit.

INTERMEDIATE BEER

In 1823, at the request of the brewers, another classification was introduced between strong and table beer. It was known as Intermediate Beer and the duty levied was 5s. per barrel. In order to prevent a repeat of the situation which arose with the introduction of table beer in the previous century, stringent restrictions were made. Intermediate Beer brewers had to be specially registered and their premises had to be at least 100 yards away from the nearest common brewer and 440 yards away from the nearest malthouse. Their casks, in order that their beer was not confused with beer produced by common brewers, had to be of different sizes and bear distinctive markings. There were also other restrictions. They were not allowed a waste allowance for the payment of duty and they had to sell the beer at a set price which gave a low profit margin. The idea was that the public should be given a low priced beer which would break the monopoly of the big brewers, but the Intermediate Beer brewers faced trading difficulties because they could only sell to the free-trade where competition was fierce. When duty was abolished in 1830, so was the classification. Only twenty-two Intermediate Beer brewers remained with a total annual production of 55,500 barrels. Out of these, eight brewed in London and had a total annual production of 51,000 barrels.

PARLIAMENT EXAMINES THE PROBLEM

Although duty was abolished, taxes were still paid on the materials used, so an incentive for diluting beer remained. In 1856 and 1874 the quality of food and drink was looked at by two Parliamentary Select Committees. In 1856 the following conclusion was reached:

> If, as regards the adulteration of articles with substances of a cheaper and innocuous character, the public derive the full benefit of this cheapness in a lower price, it would be difficult, if not unwise, for the Legislature to interfere, unless it could do so by requiring that every such article be sold as a mixture, as distinguished from the article in its pure state. Such mixtures, made known by the seller, or used for the preservation of of the article, cannot be forbidden without danger to the needful freedom of commerce,

> and ought not to be interpreted as coming within the provisions of a penal law.

In 1874 the committee again came to the conclusion that the sale of mixtures was allowable and even desirable in order to meet public requirements, providing that it was clearly indicated to the purchaser that the product was a mixture. In public houses it was common, as it is now, for people to buy drinks which are beer mixed with a weaker beer or another non-alcoholic beverage, so that the alcohol content and flavour are diluted. Also then as now, the usual and most satisfactory way of mixing drinks was to do it in the purchaser's glass so that he could see the mixture being prepared. It was never a satisfactory practice from the customer's point of view to mix or dilute beers in the cellar, as he could not see what he was buying. The report acknowledged that publicans often did not mix drinks in the glass and added water to beer in the cask; but either behaving very naively or choosing to opt out of taking any action the committee stated that:

> Parliament did not desire needlessly to hamper or fetter trade, still less to interfere between the buyer and the seller with the view to regulating prices, or attempting to assist the consumer in ascertaining the real money value of any marketable commodity.

THE 1880 INLAND REVENUE ACT

Plainly it would have been difficult to enforce any legislation as there was no standard measurement of strength for beer on which samples could be judged. However, the situation changed in 1880 when the brewer, under the new Inland Revenue Act, had to declare the gravity of the beer he brewed and this was used to ascertain the amount of duty paid. The Excise had the information and the equipment to detect any subsequent dilution and this was soon put to use in order to protect the Revenue. Every gallon of water added to a cask of beer was, under the new system, being sold by the brewer as duty-free beer, so there was now an incentive to prevent dilution which had not existed earlier. The legislation which prohibited the dilution of beer by a retailer was Section 8 of the Inland Revenue Act 1885 which also prohibited the use of unspecified substances in beer.

In the year ending March 1887 there were 624 successful prosecutions under this section of the Act for the dilution of beer. In London the number of publicans prosecuted was 313, but it was felt that during that year they had been charged such high prices by the large London brewers that many had no alternative but to resort to 'dashing' or diluting their beer in order to make a living. The pressure thereafter was very much on the London brewers to reduce their prices and prevent the dilution of beer. In March 1887 they obliged by reducing the price of their fourpenny ale and porter by three shillings per barrel, about a third of their net profit on these beers. They assumed that if the publicans stopped diluting their beer that the barrelage sold would pick up and offset the loss they were taking. This did seem to have the desired effect for a time as can be seen from the statistics in Appendix 7. However, the number of convictions started to creep up again in the 1890s and in 1897 the numbers passed those for 1887.

In the twentieth century dilution has continued with a steady though not particularly large number of convictions. There have been occasions when cases have reached epidemic proportions. One such occasion was in 1965, when fourteen licensees of houses owned by Chester's Brewery in Manchester were prosecuted for diluting their beer. Chester's Best Mild, as brewed at the Ardwick Brewery, was dark and drank very well for its gravity (1034); it could withstand a little dilution without it being detected. Several licensees were taking advantage of the fact and one of those prosecuted had added five and a half gallons to a barrel (36 gallons). A so-called mystery man reported the frauds to H. M. Customs & Excise with details of the amounts involved; it was believed that he had access to analytical equipment.

COVERING TRACKS

The dishonest retailer who dilutes his beer has the advantage that he can cover his tracks to a certain extent by adding a little household sugar to the beer. This disguises any dilution when an analysis is made, but unless a careful calculation is made it is possible to overdo the addition and give a result which is higher than that at which the beer was brewed. This is just as bad from the detecting authority's view as if dilution had been detected.

THE TRADING STANDARDS DEPARTMENT

H. M. Customs & Excise have until recent times been the authority which has detected and prosecuted offending retailers; but now, though they still have the authority, the work is mainly carried out by the Trading Standards Departments of District Councils. Brewers are also keen to detect and stop any such fraud, as it affects their sales and their reputation. The introduction of keg and tank beer has made it more difficult to dilute beer, so the number of offences have dropped, but the practice still continues. In December 1987 a survey of the licensed trade was compiled by The Institute of Trading Standards Administration.

About 3% of metered beers sampled and analysed were found to have been adulterated. In November 1987 CAMRA published the results of a small survey conducted in the West Midlands. The original gravity of nine beer samples taken showed two of them appeared to have been diluted by 10%.

WASTE BEER

EFFECT ON BEER QUALITY

The quality of beer today is higher than it has been at any other time in the past. This can be partly attributed to the way in which the brewer and the retailer now deal with waste beer. Waste beer includes that which has been recovered from the yeasty bottoms of casks, tanks and vats, beer recovered from the skimmings of fermentations, beer which has become old and stale, beer which has been spoilt by micro-organisms, beer which has been left in dispense pipes and beer from drip trays at dispense points. None of the beer mentioned above would be acceptable to a drinker by itself, so the temptation has always been to blend it off with new beer rather than throw it away. The blending off of some waste beer, such as that recovered from yeast skimmings, can be reasonably accepted but the addition of waste beer from some other sources represents adulteration of beer at its worst.

The quality of waste beer is usually low for a number of reasons. If it has been in contact with yeast cells which have stopped growing and started to autolyse, the contents of the ruptured cells give the beer a harsh flavour and raises the pH. If the beer has been in contact with air it will have become oxidised, which adversely affects flavour and assists the development of aerobic spoilage bacteria. Beer which has been kept at an elevated temperature for any length of time will be adversely affected, as temperature accelerates the rate of oxidation, the rate of yeast autolysis and the rate of growth of any spoilage micro-organisms present. A good example of this is beer which has been left in dispense pipes overnight.

References to the handling of waste beer are not very common in technical brewing literature. Where they do occur they are usually condemnations of certain practices rather than accounts of methods of dealing with waste beer. Brewers and retailers tend to develop their own way of handling the situation and establish a balance between the money lost in throwing away beer and the money lost by falling sales when they reduce the quality of sound beer by adding waste beer. It is very easy to descend into a spiralling situation where waste beer is added to sound beer and produces more waste beer to blend off. Eventually all the beer is spoilt and a good reputation is lost along with trade.

AN EARLY REFERENCE

In the *London and Country Brewer* written in the eighteenth century the author records the case of a brewster who mixed the bottoms from her casks with the the wort in the copper in order to save 'upwards of £20 per year on malt'. Her beer was held in high regard by her customers, but the author described it as a 'filthy, abominable and mischievous habit'.

PORTER

When porter brewing became established in the eighteenth century it was initially matured for up to a year in butts, but as production increased it was more convenient to mature it in large wooden vats. This permitted a characteristic stale flavour to develop. Initially the preference was for the stale flavour but later a liking grew amongst customers for the milder, newer beer. A publican was therefore supplied with both types and blended them according to the taste the individual customers required.

Returned beer or waste beer in the brewery was dealt with by collecting it together and putting it into a vat in which any sediment settled out and the beer was allowed to stabilise. It was sometimes known as old ale and was disposed of by blending it with the stale porter. The situation regarding the use of this beer, was recorded in the minutes of a House of Commons Committee, which was investigating the adulteration of beer in the Metropolis in 1818. The questioning of Mr Barclay of Barclay, Perkins & Co was as follows:

> Is sour or stale beer used in your vats with new beer, to your knowledge? – To answer the question correctly, I should state, that every publican has two sorts of beer sent to him, and he orders a proportion of each as he wants them; the one is called mild beer, which is beer brewed and sent out exactly as it is brewed; the other is called entire, and that beer consists of some brewed expressly for the purpose of keeping: it likewise contains a proportion of returns from publicans; likewise the beer which we receive from public houses, which has been brewed by other brewers, and which have changed into our trade (as it is our plan always to clear the cellar of a publican before he begins to draw our beer); and likewise a portion of the beer the bottoms of vats; the beer that is drawn off from the pipes which convey the beer from one vat to another, and from one part of the premises to another; this beer is collected and put into vats; it also contains a pro-

VATS OF PORTER

portion of brown stout, which is twenty shillings a barrel dearer than common beer: it also contains some bottling beer, which is ten shillings a barrel dearer. I should observe, that the beer returned from the publican is always examined by a class of clerks called coopers, and, as far as they can possibly judge, if there is any admixture of any kind or sort, if it has been weakened, it is put aside, and in some instances been thrown away, and the person not allowed for it; but, in general there is an examination made of the beer upon being returned. Now all these beers united are put into vats; and it depends upon various circumstances how long they remain in those vats before they become perfectly bright; when it becomes bright it is sent out to the publicans for their entire beer, and there is sometimes a small quantity of mild beer mixed with it.

What proportion of the whole number of barrels sent out would those remnants form? —About one tenth: we send out about one tenth of entire, but that is not consisting of remnants, because, I believe I stated before, that part of it is beer brewed and kept for that purpose.

Is the beer that is composed of remnants wholesome and good liquor? —Perfectly so.

Is it absolutely necessary that a publican should have some of these remnants to mix it for the taste of his customers? —I should think so. It has been the constant practice as long as I have known the trade; and in former years they used to draw more of that entire than they do now.

Does the use of stale beer effect a quick sale in the trade, and consequently a quicker return? —I do not see how the publican could well please his customers unless he had the means of making his beer either stale or mild as they wish for it. The Committee will see that if the brewer had not this vent for selling his return beer, the price of beer must be considerably higher if he is to throw this beer away, which amounts altogether to near 20,000 barrels in our house alone.

DISPOSAL OF WASTE BEER IN THE BREWERY

It is perhaps not surprising that the popularity of stale porter decreased. When the consumption of stale porter became very low and eventually stopped, the brewers had to turn to the mild ale and some of the stronger dark beers for the disposal of their waste beer. It was added either to the fermenting vessel or, if added to stronger stock beers which were matured, it was added at the beginning of maturation. Whichever method was used it was a risky business as a whole brew could easily be spoiled. If waste beer was blended off, it first had to be 'brought round', which meant that it had to be brought to a condition where it was fermenting slightly. This purged it of air. If the beer was acidic a neutralising agent may have been added, but these were not favoured by everyone.

Waste beer could not be blended off in large amounts in mild ale without affecting flavour. This often left brewers with surplus waste beer, so alternative means of disposal had to be found. One means of disposal since the use of finings had been introduced, was to cut the isinglass used in its manufacture with sour beer. However, this practice virtually stopped by the end of the nineteenth century because the risk of infecting the beer being fined was too great. Pure acids were then used which produced white finings as opposed to brown finings produced by the sour beer.

Some waste beer was sold at a nominal price to brewers in agricultural districts to sell to workers as harvest beer. This was acceptable to workers in cider-drinking areas, who wanted something different from the cider which they were given by the farmers, but were accustomed to a rough acidic-tasting drink. Waste beer was also sold to brewers in mining districts who used it to blend with their own cheap beers.

Other small brewers would buy waste Burton and Scotch ale to make into Old Beer. A recipe used in 1880 by a small brewery was to take some of this ale which sold at 20s. to 25s. per barrel and to add 3 gallons of hot first-wort taken from the hop back, then a little less than 1 gallon of stout and a gallon of hops from the first-wort. It was all put into a cask, bunged up and rolled. If necessary it was allowed to to vent before being sent out for immediate use.

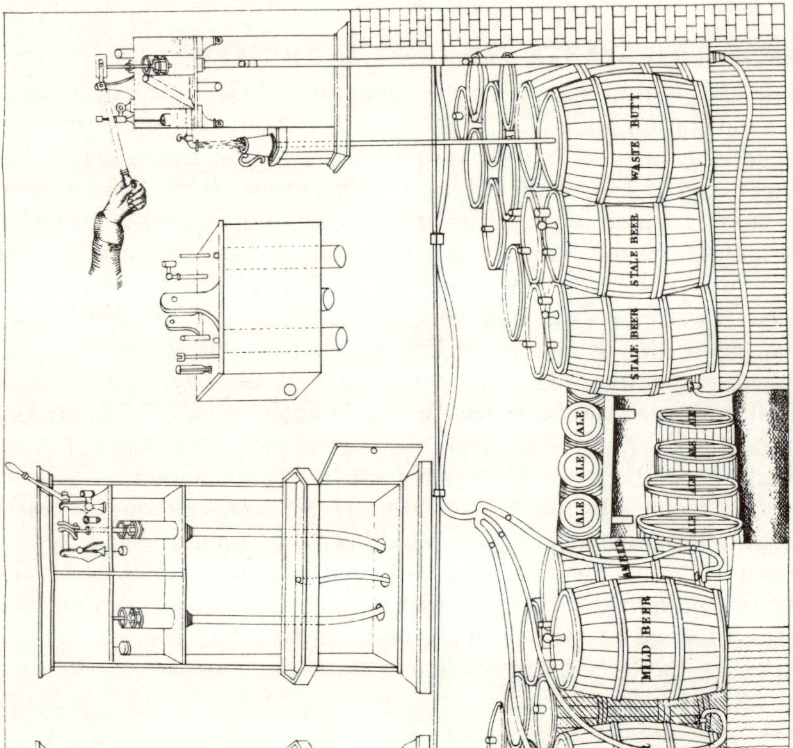

WASTE BEER IN THE PUBLIC HOUSE CELLAR 1800

As technology and scientific knowledge improved, brewers adopted better practices. Following the work of Pasteur, the importance of sterilising the waste beer which was blended with sound beer was realised. This was done by boiling or pasteurisation. Filtration and conditioning of beer also allowed it to be blended off in better condition than before. Stouts, brown ales and mild were the favourite beers for disposing of waste beer as the darker the beer the more easily any visual imperfections were hidden. Their sweet full flavour also masked flavour imperfections, which were brought about by the practice, more easily than did the less robust pale beers. This must, however, have contributed to the general turning away from darker beers. Those which have remained successful have been from breweries who have employed high standards of quality control and more advanced brewing technology.

Brewers have had the option since the reintroduction of duty to destroy spoilt beer in the presence of H. M. Customs & Excise and reclaim the duty. This is a course which many have adopted.

WASTE BEER IN THE PUBLIC HOUSE

One of the biggest problems affecting beer quality, which for various reasons has decreased in the last twenty or so years, is the practice of retailers adding waste beer to casks and even kegs of sound beer. A common piece of equipment in the cellar of a public house used to be the filter funnel. Instead of being the usual conical shape of a funnel the bottom of the body was flat and fitted with a perforated false bottom. This supported a filter paper through which the beer was filtered back into the cask. The funnel could hold several pints of beer so that all the waste could be put in the funnel at once and then left to slowly pass through the filter and into the cask. Instead of having one opening at the end of the funnel stem, this was blanked off and several smaller holes were drilled into the side of the stem, so that the returned beer was distributed evenly throughout the volume of the beer already in the cask, thus avoiding the need for any mixing. When introduced their use was seen as an advancement over the use of an ordinary funnel; brewers recommended them and gave licensees instructions as to their proper use. However, they were eventually recognised for the danger which they are and their use is generally discouraged.

One of the biggest sources of waste beer in the public house, particularly in areas of the North where a big head is expected on a glass of beer, is the overspill in the drip tray at the dispense point. A neat invention was the Autovac which drew the contents of the tray into the beer line when beer was being dispensed. In most areas the health authorities have banned their use as it is a potential means by which infection can be passed from re-used glasses and from the fingers of bar staff. However, when introduced it was an advancement over collecting waste beer and returning it to the cask. An alternative solution, though not liked by everyone, is the use of meters and lined glasses, but it is one which offers the customer the opportunity of drinking beer which has not been adulterated by returned beer.

PUBLIC HOUSE WASTE BEER FILTRATION SYSTEM IN USE IN THE 1920S

The other source of waste beer in the retail outlet is the beer which is left in the dispense lines between serving sessions. If the lines pass through a part of the building which is warm, the beer can become of unsaleable quality quite quickly and has to be drawn off. This has been overcome to a certain extent in recent times by the introduction of smaller bore lines and the cooling of lines with refrigeration.

Since the introduction of keg and tank beer, even if licensees have wished to return beer, they have not had the opportunity so easily presented to them as they have had in the past. Keg and tank beer has also cut down the likelihood of there being any spoilt beer as there is no sediment to disturb, and air cannot get into the container to cause the problems already mentioned. The keg fitting does not make the contents totally secure but like any security device it does impede the progress of the villain and deters the less determined opportunists. Cask-conditioned beer is still open to adulteration, which sometimes means that if it is being adulterated it receives waste beer from other sources as well. One method brewers have adopted in order to make retailers more responsible in returning waste beer has been to refuse credit for returns unless they are the result of a problem known to have originated at the brewery. This has met with considerable success by some breweries who have tried it.

CHAPTER 4

ADDITIVES FROM 1700 TO 1900

NEUTRALISERS

In the eighteenth and nineteenth centuries beer frequently turned sour because it became infected with lactic acid bacteria or acetic acid bacteria. Until the work of Pasteur, the famous French microbiologist, at the end of the nineteenth century, brewers were unaware of exactly what caused sourness. They realised that cleanliness was essential but technology had not developed sufficiently to stop sourness occurring regularly, especially in warm weather. The main means of prevention was to brew the beer much stronger than it is brewed today and to use plenty of hops. The best cure available at the time if beer did become acidic was to neutralise the acid. Although it removed the sharpness of the acid, it was replaced with the bitter taste of the salt produced. The neutraliser was sometimes added to beer as a matter of course so that it could neutralise acid as it was produced.

The usual neutraliser used was calcium carbonate in one form or another. Certain forms were preferred for their purity and included oyster shells, crabs claws, marble dust and egg shells.

Oyster shells and crab claws had to be specially prepared for use in beer. The usual procedure for preparing them was first to wash and dry them and then to 'bake them two or three times in an oven after bread is drawn'. Alternatively they could be left in the heat of the sun if it was summer. The brown parts were then beaten off and they were broken into pieces about as big as a shilling or powdered in order to make them ready for use. Oyster shells may seem to be a rather exotic form of calcium carbonate to use, but in the eighteenth century oysters were a common food sold in public houses. Samuel Pepys had trouble with people throwing the empty shells over his garden wall, much as people have trouble with fish-and-chip-papers today. Oyster shells were said to produce the least disagreeable flavour and easily dissolved in beer because of their nature.

Marble dust was obtained from sculptors and ground to a consistency of coarse sand. The great advantage that it had in use was that, because it was hard, it reacted with the acid very slowly and the carbon dioxide produced was not released in a gush. Quicker acting neutalisers caused problems with the beer fobbing and loss of most of the gas. With a slow release of carbon dioxide gas much more of it dissolved in the beer and the sharp flavour it gave masked some of the more undesirable flavours present.

Egg shells had to be washed and baked before use. Whole eggs were sometimes added to beer instead of just the shells. Dr Stubbs writing in the eighteenth century tells of the practice which he learnt from an ale-seller in Deal, who believed that it had been practised for centuries. It may have been that the egg white which is known to have a fining action was also intended to play a roll in fining the beer.

Alabaster is another exotic form of calcium carbonate which was used to neutralise beer, but since there are two forms one of which is calcium sulphate and the other calcium carbonate, care was needed to select the correct type.

The more common forms of calcium carbonate, limestone and soft chalk, were also used but not considered to be so good. They were often added with the expectation of providing some fining action as well; other particulate substances such as sand were sometimes added for the same reason.

Neutralisers were often added in conjunction with other substances in order to serve a multiple function. They were sometimes added to beer with molasses, honey or treacle if the beer was flat as well as sour. The sugar restarted fermentation and also masked some of the unpleasant flavours.

Caustic alkalis were tried as neutralisers but also reacted with the carbon dioxide in the beer and made it quite flat. In addition the flavour they produced in the beer was quite nauseous. Quicklime was also found to be an unsuitable neutraliser because of the flavour it produced, though it was recommended occasionally for forwarding a fermentation because of its heating properties.

HEADING AGENTS

One of the characteristics of porter was its thick frothy head and in order to ensure that it had a good one, it was common practice to add substances known as heading. Initially it was the brewer who added the heading at the brewery, but eventually it became practice for the publicans to add it. This was presumably because the brewers were persuaded to stop using it whereas some publicans chose to continue the

NO MORE FLAT OR STALE BEER.

EXPOSITION UNIVERSELLE, PARIS, 1878.

FRENCH CREAM.

GUM EXTRACT.

(Protected by Royal Letters Patent.)

FOR PRODUCING A PERMANENT HEAD OF CREAMY RICHNESS UPON ALL KINDS OF ALE AND BEER.

As a brisk and sparkling appearance adds so much to the quality of Beer, it is invaluable, and should be put into every cask, and no **Bottled Beer** that is wanted to **open with a fine sparkling head** should be without it.

This will be maintained, and **improve greatly the condition of Beer** that would otherwise become stale or flat.

It is perfectly harmless, does not promote acetous fermentation, has no deteriorating action whatever on the keeping quality of the Beers; on the contrary, its antiseptic properties conduce to their preservation. It should be used in high-class Ales, Stout, and Porter, as well as in lower qualities.

To be used in the proportion of four ounces to a Barrel—36 gallons— and added just previous to the racking for delivery.

In 1 Gallon Bottles, 2s. 6d. per lb.; in Quarter Casks, 2s. 3d. per lb.

GUM EXTRACT IN POWDER.

A preparation of the above in a solid form specially adapted for Beer and Ales Brewed for Exportation.

In 1 lb. Bottles, 18s. per lb.; in 7 lb. Bottles, 14s. per lb.

CAUTION.—All persons infringing the above Patent will be proceeded against. The only genuine Gum Extract, Fluid and in Powder, is that prepared by Messrs. W. J. BUSH & Co.

MANUFACTURED ONLY BY

W. J. BUSH & Co., 20 to 23, Artillery Lane, LONDON.

Works:—ASH GROVE, HACKNEY.

practice. The main substance used was iron sulphate, known as green copperas, green vitriol or salt of steel. It was used alone or mixed half and half with alum which is also a heading agent but less effective. Sometimes common salt was also added. The amount of heading recommended for use varied between ⅛ oz/brl and ½ oz/brl. A common guide for measuring the amount required per barrel was to put in as much as would lie on a shilling.

One of the consequences of its use was that when porter, to which it was commonly added, was drunk out of a pewter tankard or tin pot it gave it a peculiar flavour. This was said to be favoured by porter drinkers, though it is doubtful if the chemical produced by the heading reacting with the pewter did them much good. Publicans approved as the chemical also gave the drinker a thirst.

It is of interest to mention at this point that salt of steel was also added to the mash at the rate of 2 oz per 20 barrels in some parts of the country. The idea was that it scavenged oxygen from the water and prevented the growth of acid producing bacteria. They did not know at this time that lactic acid bacteria which cause acidity of the mash are anaerobic. It was also said by others to increase the extract from the malt.

It appears that salt of steel and green vitriol or copperas, although chemically the same substance, were different in some way as one source states that less salt of steel was required than copperas or green vitriol.

The use of these substances was illegal and attempts were made to eliminate their use in the early part of the nineteenth century. The heading agents which were widely advertised at the end of the nineteenth century were a different substance altogether. It was quillaia bark which was powdered or made into an extract. The powder was often mixed with magnesia to give it bulk and to keep it dry, and sulphites were added as preservatives.

One of the crudest adulterants which was used to produce a head, and was even mentioned in Shakespear's *Henry IV*, was soap. It was believed to have been used in the eighteenth and nineteenth centuries but must have had limited appeal.

PRESERVATIVES

Very few food preservatives had been discovered by the eighteenth and nineteenth centuries so there was little misuse of them. The most widely used preservatives, as they are even today, were the salts of sulphur dioxide. One of their first mentions in connection with brewing is when

PRESERVATION OF ALCOHOLIC LIQUIDS.

PRENTICE BROTHERS, STOWMARKET,

ORIGINAL MANUFACTURERS OF

D C S

TRADE MARK.

DRY CALCIC SULPHITE.

Proved to be particularly adapted for the preservation of all Fermenting Liquids. Attention is drawn to its demonstrated success as a restorative of sour beer to its original brilliancy and soundness, with absence of returning acidity. The best material for cleansing and rendering old brewing, fermenting, and distilling plant perfectly sweet.

"CUM GRANO SALIS."

CUM GRANO SALIS

TRADE MARK

The FIRST and ORIGINAL PREPARATION for Soft and Impure Waters, rendering them suitable for the production of Sound and Brilliant Ales.

Manufacturers of SOUTHBY'S PURE CALCIC CHLORIDE SOLUTION for imparting palate fulness to beers. A and N solution for treatment of waste and other preparations. For further particulars apply to—

GEORGE A. CLOWES & CO.,

SOLE AGENTS,

NEEDHAM MARKET.

Henry VIII wrote to his brewer in Eltham in 1530 and told him to stop using hops and brimstone in his beer. The brimstone would have been burnt to produce sulphur dioxide which acted as a sterilant or preservative. One way in which it was used was to burn it inside casks just before they were filled. Another use, which was only discontinued a few years ago, was to burn sulphur in the kiln when hops were being dried. This may have been the use to which Henry VIII was referring, as hops and brimstone are mentioned together.

The use of oil of sulphur now known as sulphurous acid was mentioned in an eighteenth century book, *The London and Country Brewer*, but the use of specially manufactured salts are not recorded in brewing until the end of the nineteenth century. Then, sulphite of soda was recorded as being better than sulphite of lime and both were more effective than sulphurous acid. It was considered that there was not a brewery in the country which was not using sulphites as preservatives at the end of the nineteenth century.

The other preservative in use at the end of the nineteenth century was salicylic acid. A patent was taken out for the use of the soluble salts, in 1877, by Frederick Burbidge, a manufacturing chemist. He patented the use of the soluble salts as it was not practical to use salicylic acid itself, because of its limited solubility. The salts are ineffectual preservatives but the natural acids of the beer caused a reaction which produced salicylic acid. The salts which he said could be most easily produced were those of 'lime, magnesia and strontia'. The recommended dose of the salts was between ¼ and 1 oz per barrel for preventing infection developing and higher if acidity was being checked. The preservatives were tasteless when first added but after a few months the beer developed an odd and objectionable flavour. As a result the preservatives could not be used in stock beers but could be used in running beers which were sold with little or no maturation. It was also found to be unsuitable for the preservation of finings as it made the isinglass go hard.

Salicylic acid was said to have come into use in Britain around 1885–86 when a reputable brewers' chemist had recommended it for purifying yeast. It was reputedly used more for this purpose than as a preservative in beer. However, there was conflicting evidence on the extent of its use as a preservative at the end of the nineteenth century, when this was examined by the Parliamentary Select Committee investigating the use of brewing materials in 1897. One witness stated that its use had been discontinued in this country about ten years previously because of its effect on the flavour of beer and because it flattened the head. Another witness, however, said that he had been informed by a Continental supplier that if its use in beer was banned in Britain, that their British sales would drop by 25%. A further witness said that he believed that there were 800 brewers using it. One public analyst submitting evidence said that he had obtained convictions under the Sale of Food and Drugs Act for the use of salicylic acid in beer, which suggested that it was not regarded as wholesome and not in common use. Evidence from suppliers had to be regarded with some suspicion. They perhaps did not wish to jeopardise what few sales they did have in the U.K.

The Germans were particularly keen on the use of salicylic acid as Kolbe, a German, had discovered and patented a cheap method of producing it. In the eyes of the French this alone was enough to prejudice its use. Their fears were eventually proved to have grounds and its use in beer was banned. However, this did not stop Charles Perrot of Paris patenting a preservative in London in 1878 which included salicylic acid as an ingredient along with cream of tartar, tartaric acid and tannin.

Borax was another chemical classified as a preservative at the time, but the scarcity of details of its use indicates that it was not very popular.

FININGS

Isinglass is without doubt the most successful substance which has been used for fining beer. It was first used in British beer in the eighteenth century and is the only one which has remained in continuous use up to the present time. Its history is detailed in the next chapter with other additives used in the twentieth century.

Various herbs have been known for centuries to have a fining action on beer and were widely used as fining agents before the introduction of isinglass. The best-known herb was ground ivy which was also known as ale hoof or tun hoof because of its properties. Other herbs used for fining beer were hyssop, vervain and balm.

Minerals were also used for fining and acted by collecting yeast on their surfaces as they sank through the beer to the bottom of the cask after being added through the bung hole. Substances which are recorded as having been used include: small red gravel, scouring sand, river sand and white pebblestones which had been baked and powdered. Minerals were usually burnt in order to remove any organic material and any earthy taste. The main consideration was that they did not taint the beer. The disadvantage in their use was that they sank to the bottom of the cask too quickly and did not have chance to collect all the yeast.

Several other substances were used which were a little lighter and drifted to the bottom of the cask more slowly. Whole grains of wheat were one such substance and were considered to have the advantage that the yeast could feed on them once they had carried out their fining action. Two other substances which enjoyed some popularity, perhaps because of their exotic nature, were ivory shavings and hartshorn shavings. Hartshorn shavings were the raspings or shavings of the horns of male deer and, like ivory, they were an expensive commodity. Although these two substances were used to fine beer in cask they were also used as copper finings. They were added to the copper and boiled with the wort at the rate of 1 lb per hogshead. They dissolved and clarified the wort because of their glutinous properties. It was also believed, at the time of their use, that they preserved the beer because ammonia was distilled off and increased the alkalinity of the beer. We now know that this view was scientifically unsound. However, brewers of the time did have enough practical experience to know that if too much was added, a urine-like flavour would result and the wort would become thick.

Wood shavings were another substance which became popular for fining beer at the end of the nineteenth century, particularly with lager brewers. Beech shavings were the most popular, though hazel and cherry wood were also used. The shavings used were long and broad and not too thin. They were boiled or steamed before use in order to ensure that they were sterile and could be reused several times over if properly washed and sterilised. Special shavings-washing machines were even available to make the job easier. It was found that besides fining the beer they also made it attenuate better, because the shavings and yeast which settle out present a larger surface area of yeast cells than a compact sediment of yeast. One very large American brewery company has continued using beech shavings in their krausening tanks in which a secondary fermentation is carried out. The shavings, or chips as they are also known, are 2 ft long, 4 in. wide and ¼ in. thick and are put in the tanks to a depth of 4 ft. In the twentieth century a new product called Biochips was introduced to replace beechwood chips. It was sterilised sawdust weighted with pitch or mammut so that it did not float. Later aluminium foil strips and rings were also used.

Numerous other substances have also been used over the centuries to fine beer. A recipe for finings taken from a French work of about 1730, which even then was claimed to be old, is given as follows:

> Take three pounds of powdered nut-galls and four ounces of potash. Boil these for three hours in such a quantity of water as, at the end of that time, will make the weight of the whole mixture about twelve pounds. To this, when cool, add two pints of spirit of wine; and, after it has settled and become clear, bottle it up for use. Five ounces of this decoction will be sufficient to fine and preserve half a piece of beer.

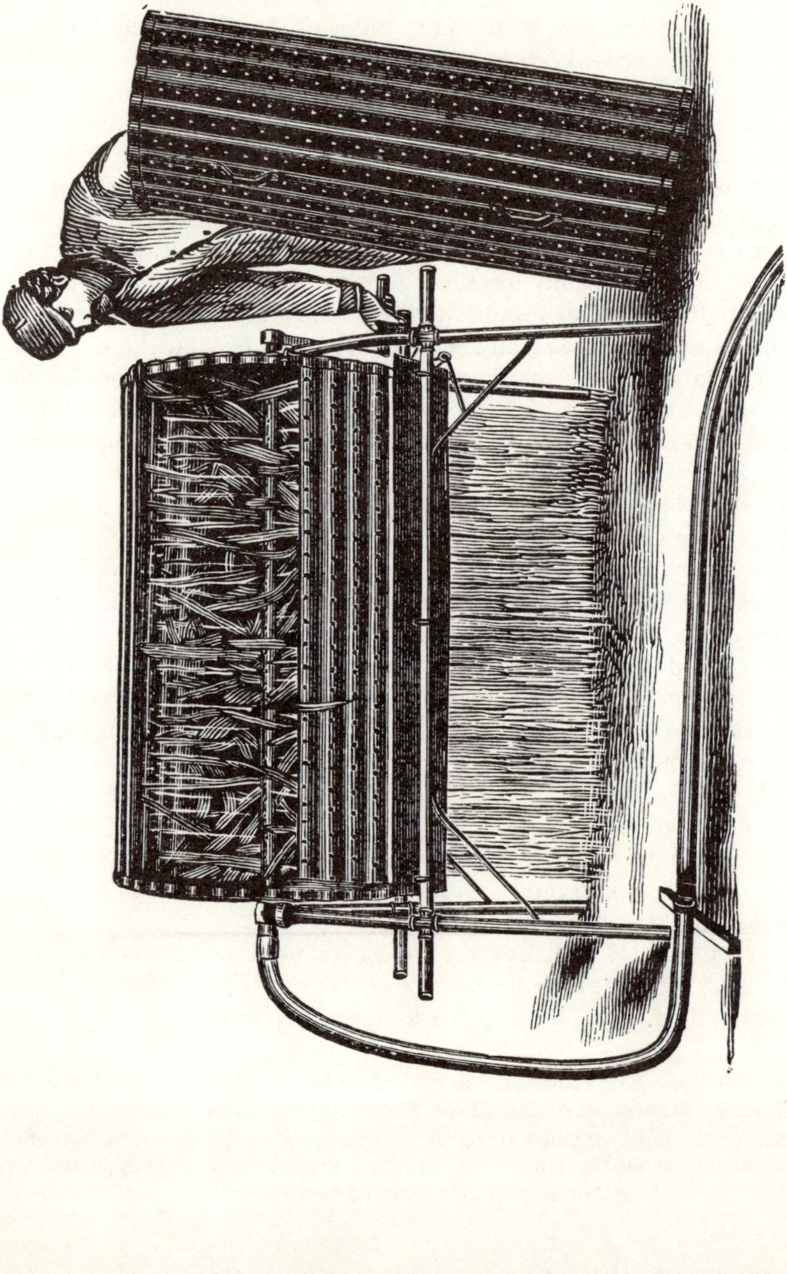

A SHAVINGS WASHING MACHINE

Some of the substances less widely used for fining in the eighteenth and nineteenth centuries included: boiled wheat, egg whites, blood serum and alum. Alum was laterly added to beer at the rate of 4 oz per hogshead before isinglass was added, in order to reduce some of the bulk of solids in suspension.

In 1886 Dr Eugene Prior of the Nuremberg Laboratory published a paper on the use of liquorice as a fining agent. Its use had apparently been well known on the Continent but the flavour it produced must have put it at a disadvantage. Though liquorice was used as a hop substitute its flavour is one that a brewer would not necessarily like to give to his beer.

Sulphuric acid is mentioned as a fining agent by one author but it solved the problem in a unique way; solids were dissolved rather than precipitated. Needless to say it was regarded as an adulterant rather than a processing aid or additive.

MISCELLANEOUS ADDITIVES

SUBSTANCES USED TO ACCELERATE FERMENTATIONS

In the eighteenth century various substances were used to accelerate, or 'forward' a fermentation as it was then described. Some of the substances added may seem to be a little strange today unless one understands what brewers saw as the reasons for a fermentation proceeding slowly.

They thought that if wort was too viscid or clammy, caused by it being bound together with oleaginous (fat-like) particles, it would ferment slowly. They considered that the particles had to be dissolved in order that the texture was broken down, for the fermentation to proceed more freely. Substances which fulfilled this purpose included salt of tartar, potash, quicklime and burning coals or wood.

Sometimes a slow fermentation was considered to be caused by poor yeast and its activity was often revived by adding flour and salt, or just flour, to the brew. The small particles of flour or salt act as foci for the evolution of gas bubbles. The release of these carry yeast, which has settled on the bottom of the fermenting vessel, back into suspension. Other granular substances were also used with flour including: saltpetre, salt prunella, pepper and ginger. This practice has continued right up to present times and is now known as dressing a fermentation. Some brewers dress fermentations at a certain stage as a matter of course and use sterilised inert materials such as kieselghur as well as flour and salt. A typical rate of addition is 1 to 2 oz of flour and ½ to 1 oz of salt per barrel. On some occasions dressing is also used at the end of fermentation to help lift the yeast up to the surface to form a thick head which is cropped off.

It was also recognised in the eighteenth century that a fermentation could stop because of a lack of nutrients. Sugar, treacle and honey were all used to revive languid fermentations, though because of Customs and Excise restrictions it was not permitted after 1880 without formal notification.

SUBSTANCES USED TO CHECK A FAST FERMENTATION

At the present time with refrigeration and other technology it is an easy matter to control the temperature of beer, but in the past fermentations could race away and beer fret excessively in the cask in warm weather. In the eighteenth century an active fermentation was seen as 'abounding with subtle [elusive] and fugitive [unstable] particles'. The cure was to check the activity of these particles, which we now know as yeast, with various inhibitory substances. A number of chemicals were suggested including salt, alum, nitre, spirit of vitriol, oil of sulphur, spirit of salt and other acids. The burning of brimstone under the fermenting vessel, which would have produced sulphur dioxide gas, was another remedy. The addition of cherry brandy was seen as a means of checking the fretting of a maturing beer but it must have been added in large amounts for the alcohol to inhibit the activity of yeast.

SUBSTANCES USED TO IMPROVE FLAVOUR

In order to save on the cost of maturing porter, freshly brewed mild porter could be given a smack of age by adding alum which has an astringent taste. It was added in the maturation vat, sometimes alone and sometimes with salt. Hydrochloric acid was another chemical used for the same purpose. Ten to twelve drops were put into a quart of ale. Sulphuric acid is quoted by another source to have been used to improve the heating qualities of beer.

It was recorded in the eighteenth century from information supplied by a brewer, that it was common practice for human urine or chamberlye to be added to pale or twopenny amber beer, in order to improve the flavour. The author described it as a nasty, horrid and detestable piece of cunning and knavery.

SUBSTANCES USED TO EXCLUDE AIR FROM BEER

It was recognised in the eighteenth and nineteenth centuries that it was desirable to keep air from coming into contact with beer in cask or bottle. An oil, such as olive oil, was sometimes put into bottles of beer so that it floated on the surface and formed a barrier. When the beer was required the oil was removed by soaking it up with a piece of cotton. The oil would have destroyed any potential for producing a head on the beer but presumably this was of low priority compared with preserving flavour. Powdered rosin was sometimes added to cask beer when it was dry hopped; both substances floated on the surface of the beer and formed a seal.

SUBSTANCES USED TO CURE TAINTED BEER

In the eighteenth and early nineteenth century the likelihood of beers becoming tainted through microbial spoilage was very high. The situation was such that brewers were inclined to try to remove the taint rather than destroy the beer and start again, suggesting that there was an inevitability that taints would be produced. This was most likely due to inadequacies in the equipment used for brewing which prohibited it being properly cleaned and because temperatures could not be controlled in hot weather.

The common approach to the problem was to cure it in the cask by adding some substance which absorbed the flavour of the taint. Starch in some form was the usual remedy and the substances used included wheat, malted oat, pea or bean flour which eventually sank to the bottom of the cask. Whole wheat or horse beans were also used and were often hung in the beer in a bag, but horse beans had to used with care as they could give the beer a tang if used in too great an amount. Toast was sometimes put into beer and punch when it was served for the same purpose. A French observer said that the English were right for doing this but fools for eating the toast.

Another substance used was beech ash which again was usually hung in a bag in the beer either in the fermenting tun or in the cask.

Recipes for some other cures contained alum as a common ingredient. One recipe included 1 oz alum, 2 oz bruised mustard seed, and 1 oz rasped or bruised ginger. It was enough to cure half a hogshead of foxed or ropy beer. Another recipe contained alum, mustard, treacle and beanflour and a third one contained alum and horsebean flour.

As has been mentioned, the quality of malt was often poor and could carry taints, particularly from smoke. An old practice used to cure taints from malt was to add a shovelful or two of hot coals to the mash. However, it was considered better practice to quench a firebrand or a burning cinder in the wort so that the grains were not spoilt. Another practice was to put burning beech billets, burnt almost to a coal, into the copper and boil them with the wort. Bizarre though this may seem, it perhaps worked, as charcoal is used in sugar refining to remove unwanted flavours and even in brewing to remove flavours from waste beer which has been processed.

SUBSTANCES USED TO IMPROVE HOPS

It was common practice in the eighteenth and nineteenth centuries to rub hops with salt of tartar. *The London and Country Brewer* claimed its use as:

> . . . a new method of great service in forcing the oily part of the hop to emit its viscid quality, and expeditiously join in contact with that of the wort: for by this the floury part of it may be obtained, clear of the earthly phlegmatic, which long boiling never fails of extracting, to the great prejudice of the beer and the drinker's health.

A recipe for a brew using 3 lb of hops required ¼ oz of salt of tartar.

DUMPLINGS OR FINING BALLS

These were fashionable additions put into casks of maturing beer in the eighteenth century. They served several purposes, namely to fine, feed, preserve and to relish a beer so that the beer at the end of maturation was bright, sound and well flavoured.

Their basic ingredient was usually flour; malted barley, wheat, rye and bean flour were all recommended. The preservative ingredient was usually a neutraliser which was regarded as a preservative in those days as it removed acidity. Chalk, oyster shells and other sources of calcium carbonate were used. Some sugar source such as treacle or molasses was included and can be regarded as an early form of primings. Isinglass was also sometimes added. Various herbs were used as flavouring and frequently included hops which was perhaps the origin of dry hopping.

All these ingredients were mixed together into balls which could be put into the bung hole of the cask. Eggs were often included in the mixture in order to bind everything together as well as give some fining action. If the shell was included it also acted as a neutraliser.

An example of a recipe is taken from *The London and Country Brewer*:

> Alabaster, or marble calcined into a powder, two pounds. Oyster-shells a little calcined and freed from their brown or dirt coloured outside, one pound. Pure fat chalk, well dried, one pound. Horse bean flour, first freed from the hulls, one pound. Red saunders, four ounces. Grains of paradise, half an ounce. Florentine orice root, half an ounce. Coriander-seed, a quarter of an ounce. Cloves, in number, six. Hops, half an ounce. The best staple

> incised isinglass, two ounces. The first runnings of the molasses, or treacle, two pounds.

The ingredients were made into balls weighing 4 oz each, and three were put into a barrel. In this case the balls were also intended to give the beer colour which came from the red saunders and the molasses. If required the red saunders could be left out and the molasses replaced by sugar.

HOP SUBSTITUTES

REPEAL OF THE HOP DUTY

On 16 September 1862 the hop duty was repealed after a campaign by the hop growers, who had even gone to the extent of paying an agitator to further their cause. The relief of this burden was soon replaced by another, so their delight was short lived. The reason was that when duty was paid on hops, the Excise were very active in suppressing the use of any substitute which would result in a loss of revenue, but with the repeal of the duty it became legal to use hop substitutes as long as they were not malt substitutes as well. The Act repealing the duty certainly did the brewers no good, as the duty on their licences was increased to cover the loss of the revenue from the hop duty. The price of hops did not fall and by 1875 it had increased by 20%. The hop growers also gained nothing and faced the additional competition from hop substitutes. Some hop growers, realising the mistake, agitated for the reintroduction of the hop duty; but of course they received no support.

THE FIGHT AGAINST HOP SUBSTITUTES

Eventually a meeting was called in February 1875, in Maidstone, by the hop growers and was attended by about a hundred people. There was clearly some concern amongst those attending that the use of substitutes would seriously affect the trade. Mr Wilmot, a hop merchant, told the meeting that he had tried some substitute and found that it had 'as nice a flavour as anything he had tasted'. He told of one brewer who had asked a gentleman 'whether he thought it likely that he should give £10 or £12 a hundredweight for hops, when he could produce a compound for 6s. or 7s. a hundredweight that would answer the same purpose'. He told another story of a gentleman who went to a brewers' chemist and was offered bottled stout produced without malt or hops, which again was described as 'as nice a glass of stout as ever he had tasted'. His call for action, after relating these stories, was well received by the meeting with cheers and cries of 'Hear! Hear!'. Mr Wilmot's proposal 'That a committee be formed to call upon the President of the Board of Trade to inquire on the questions of the use of deleterious mixtures as hop substitutes', was carried by the meeting.

It was found that Mr Sclater-Booth was jointly introducing the Sale of Food and Drugs Bill with Mr Clare Read in the current session of Parliament. The Committee decided to try to persuade him to amend the bill so that it stated that beer should only be made from malt, hops, sugar, salt, water and nothing else. They wanted the ingredients of beer defined so that prosecutions for the use of hop substitutes could be easily made. Sclater-Booth's response was, that although he wished to help the hop growers he could not weaken the Bill by including any special regulations. It was then noted that Clause 5 of the Bill stated that:

> No person shall sell, to the prejudice of the purchaser, any article of food, or any drug which is not of the nature, substance and quality of the article demanded by such purchaser . . .

and Mr Sclater-Booth was then asked to add the following:

> . . . or which shall be of any other nature, or composed of other materials, than shall be expressed by the name of such article.

or

> . . . or to be used as a substitute for the material of which such article professes to be composed.

On consideration, he declined as again he felt that it would weaken the Bill.

These events were reported to another meeting of the hop growers in April 1875, but the Committee felt that all had been done that could have been done and that the Bill, if passed by Parliament, would offer them the amount of protection they required. A resolution was made: 'That this Committee beg to express their satisfaction with the Sale of Food and Drugs Bill, as amended, and trusts that the 6th clause of the Bill will be maintained.' The sixth clause read:

> No person shall sell any compound article of food which is not composed of ingredients in accordance with the demand of the purchaser, under a penalty of twenty pounds.

However, they did feel that it would be necessary to make it widely known that when a man asks for beer, he is entitled to have a drink made from malt and hops alone. The hop growers had, if nothing else, the foresight to see that this was the strongest clause in the Bill; it turned out to be the one under which most prosecutions were made. The third clause was intended to be the main clause under which prosecutions were to be made but it was so worded that the prosecution had to show intent on the part of the adulterator, which was virtually impossible.

Unfortunately, the Act when passed did not offer the hop growers the protection that they were expecting and no prosecutions were ever made in regard to the use of hop substitutes. On paper, beer was covered by the Act which was administered by Local Authorities but in practice the substances added to beer were under the control of the Excise branch of the Inland Revenue, using their laboratories at Somerset House. The public analysts employed by the Local Authorities, though they had the authority to analyse beer and prosecute brewers under the 1875 Act, tended to leave the matter to the Inland Revenue. The Inland Revenue had great experience in analysing beer for adulteration as well as having officers present in all breweries who could observe any malpractices. They were acknowledged as the greater authority and in several cases where they had been asked to check on cases brought by public analysts, they had criticised their work. This situation was not in the best interests of the consumer as the Inland Revenue's main concern was to protect the revenue, not the consumer. As hop substitutes were no threat to the revenue they were ignored.

THE HOP FAMINE

The next blow to the hop growers was the hop famine of 1882 which was the worst one since 1816 and 1817. The large brewers were little affected as they saw the crop was going to be poor and immediately secured the bulk of the best hops, leaving the remainder to the small brewers. It was calculated that 500,000 cwt of hops would be required for the year, based on a usage of 8 lb per quarter of malt. It was assumed that if the brewers economised the amount would be reduced to 450,000 cwt. The opening stock for the season held by the merchants owing to poor trade the previous year was 120,000 cwt. Although the total weight of hops picked was low the quality was good and said to be one and a half times the value of those picked in the last three years, which improved the situation slightly. The remainder had to be made up with imported hops and the shortage raised the price of hops by £15 per cwt over the previous year leaving, it was said, the brewers to pay an extra £7 million.

A letter was published in *The Times* in November 1882 from a Mincing Lane drugs broker stating that the price of bitter drugs had suddenly risen enormously in price. These drugs included colombo root, camomile, quassia, Guinea grains and cheretta. The large brewers were quick to deny the implication that they were being used in beer. Their claims were perhaps justified as they had secured their requirements of hops quite early on. Some explanations were put forward at the time and included the observation that if drugs were being widely used the price of hops would have dropped, but it did not. Another suggestion was that the price of the drugs was being pushed up by dealers who had cornered the market and was no indication of the extent to which they were being used. It was a fact that the price of quassia had probably risen because the hop growers themselves had been using large amounts to combat mildew and lice. This is shown by its enormous price rise from £5 to £40 whereas the price rise of camomile was more modestly from £2 to £6.

PARLIAMENTARY INVESTIGATION

In 1890 a Parliamentary Committee reported on the hop industry and established the fact that hop substitutes were being used. One witness estimated that their use was 1% of the total hop usage. The recommendation of the Committee was that unless they were deleterious they could be used, but their use should be declared by the brewer. However, no further action was taken by the Government on the recommendation.

In 1897 there followed the report of another Parliamentary Committee on brewing materials, in which recommendations were also made on the use of hop substitutes. In the course of their investigations the Committee took evidence from brewers, members of the Inland Revenue, public analysts, various consultants and members of the hop trade. The brewers questioned claimed, perhaps not surprisingly, that they had never used them except in the year of the hop famine. Some even doubted their existence, whilst others admitted that they had been visited by travellers trying to sell them substitutes. However, evidence of their use was more forthcoming from other sources. The Excise had been asked to collect for the Committee some evidence of their use by brewers. A surprising number of brewers when questioned by the Excise, said that they had just stopped using them and had no beer in stock made with substitutes. One suspected hop substitute discovered in a brewery was found to be tannic acid, but it was said by the brewer to be used to precipitate undissolved starch in the cooler and did not pass into the wort. Another one discovered was commercial catechu used to precipitate 'extra albuminous matter' and again, since it formed a precipitate which did not pass into the beer, it was not considered an additive. Only one sample discovered was a hop substitute and was identified as quassia. It was said to be used in a coarse beer which had a local popularity and suited the taste of the customers. About forty brewers out of 8,843 were registered as using hop substitutes including one large brewer. Mr Bannister, late of the Inland Revenue Laboratory, submitted written evidence that in the past the following substances had been used as hop substitutes at different times: quassia, chiretta, calumba root and gentian.

Other witnesses produced strong evidence on the use of substitutes, which supported the facts given by the Excise. In addition they testified that camomiles and picric acid were used. *Cocculus indicus* was said not to have been used in recent times but that its use was known of in Russia. Evidence was also presented which showed that hop substitutes had been well advertised since the hop famine in 1882 and were still being advertised in 1897. There seemed to be no doubt that hop substitutes had been used in the hop famine of 1882 and had continued to be used after that date, probably to save money and perhaps because customers had developed a taste for them. They were used by both large and small brewers, but more commonly by the small public house brewers in the North. They seemed to be popular in certain cities and towns, Manchester being named as one city where they were widely used. The use of substitutes was, however, seen by the committee to be on the decline perhaps because the price of hops had become more stable. This was attributed to the fact that hops could be kept for longer periods using cold storage and there was a larger buffer stock which could be drawn on in times of shortage.

Eventually the Committee managed to secure evidence from the main manufacturer of hop substitute, Nathanial Bradley, who produced 'Hop Supplement'. He said that he had a virtual monopoly of the trade, particularly in the North of England, having fought off many competitors and imitators. His product was also used in herb beers which were particularly popular in the North. Through fear of imitation he refused to reveal the composition of his product, but would say that it did not contain *Cocculus indicus*. He claimed that it enabled the brewer to obtain consistent bitterness levels and that in bad years they did not have to resort to the use of poor hops. He also said that the product could be used without hops, providing that the beer was dry hopped, but it was rarely used to replace more than a third of the hops. He sold not only to the public house brewer but also to the larger brewers. Of the 603 brewers brewing over 10,000 barrels per annum 1% or 2% used his hop supplement. He claimed that even some Burton brewers, who staunchly denied using hop substitutes, had used it in times of shortage.

After considering all the evidence the Committee decided that no deleterious substances were being used as hop substitutes and that no

TWENTY-FIRST YEAR OF CONTINUOUS SUCCESS.

See our Registered Trade Marks on all Articles supplied by us.

Our Customers have obtained PRIZE MEDALS in England, America, Germany, and South Africa.

BRADLEY'S APPARATUS FOR PURIFYING PITCHING YEAST.

Testimonials from Brewers who make sound, vigorous Yeast without change. The only Machine and method of purification protected by Royal Letters Patent.

"The Hop Supplement."

(Called "H.S.")

To save one-third of the Hops, and supplement the Hops used with valuable keeping properties. The "H.S." produces a more delicate flavour in the finest Pale Ales.

By Post, Packet equivalent to 4 lbs. Best Hops, 1s.; 10 lbs. 2s. 6d.; 50 lbs. 12s. 6d.
Sold in Packets of all Sizes, Carriage Free.

PRIVATE TESTIMONIALS FROM HIGH-CLASS BREWERS WHO HAVE USED IT SUCCESSFULLY FOR SEVENTEEN YEARS.

NO MORE FLAT BEER.

Bradley's Patent Dry "Frothing Powder."

Dissolves bright, and produces a good lasting head on Ale and Porter. Sample by post 1½d., to cream 36 gallons to the glass. Gives neither taste nor smell. Price 8s. per lb. Large quantities, special terms.

Bradley's Beer Preservative.

(Called "B.P.")

Unequalled. Costs less than ½d. per barrel. Is a dry powder, without smell or taste.

Bradley's Permanent Hardening.

(Called "P.H.")

For rendering the Mashing Liquor equal to the celebrated Burton Springs. Readily Soluble. Twice the available strength of any other Burton Water Material in the market. In cwt. kegs, 30s. per cwt.

Pasteur's "Yeast Nourishment."

(Registered "Y.N.")

This is the Preservative Yeast Food which is approved and extensively used in England and Abroad. 10 oz. packet sufficient for 20 barrels, 1s. In bulk £8 8s. per cwt.

Bradley's Mash Material.

(Called "M.M.")

To use in the Mash for purifying the Wort from the injurious portions of the Malt, for protecting the Wort from acidity, and for producing a Beer of greater delicacy of flavour and colour, even from inferior Malt. Guaranteed of highest purity and greatest strength. 1s. per lb.

MANUFACTURED ONLY BY

NATHL. BRADLEY, F.C.S.

(*Proprietor of the Old Firm of* ESTCOURT & CO.)

49, 51, 53, 55, Chapman Street, Hulme, MANCHESTER.

Analysis of all Materials connected with Brewing by NATHL. BRADLEY, F.C.S., Analyst to the Manchester Brewers' Association.

TELEGRAPHIC ADDRESS—"ANALYSES, MANCHESTER."

useful purpose could be served by enforcing their declaration when beer was sold. It was suggested that a line be included in the *Excise Brewing Book* so that the brewer had to show the amount used in a brew. This would, it was said, help the Excise keep an eye on the levels of use and ensure that no deleterious substances were used. This suggestion was carried out and each year the amounts used were published in the Annual Report of the Board of Customs and Excise. This continued until 1959 when the total annual usage had fallen to 29 cwt.

THE NATURE OF HOP SUBSTITUTES

The composition of a couple of hop substitutes can be found in patent records. In 1856 Richard Brooming of Fleet Street, London, patented a substitute which originated in France. The compound was known as 'Lupuleid' and made from a resinous substance, to which nitric acid was added, preference of use being given to white Burgundy pitch. In manufacture the mixture was gently heated until the resin started to bubble; it was then removed from the heat and the resin allowed to bubble over the side of its container to be caught in a tray. The resin was then cooled, washed to remove traces of acid, dried and powdered. It was supplemented by the bitter principle from aloes and 1 lb of the finished product used to replace 10 lb of hops. In 1877 Harry Houlden Terrins of Handsworth, a sweet dealer, took out a patent on a hop substitute which was basically aqueous extracts of quassia and gentian mixed with an alcoholic extract of hops and some caramel.

Technical brewing literature of the time gives a little indication of the nature of hop substitutes. Faulkner, in 1884, recounted that there were many of them and that they consisted of some bitter principle such as quassia, chiretta or gentian usually mixed with some resinous matter and a little tannic acid. He did not condemn their use but could find nothing to commend it. Southby, in 1885, goes a little further and recommends amounts of certain substances as suitable replacements. One pound of the following substitutes was said to be equivalent to the following number of pounds of hops:

Quassia	16 lb
Calyso	12 lb
Chiretta	10 lb
Gentian	7 lb
Camomile	5 lb

He considered these substitutes to be wholesome but stated that their use was undesirable, that they should not be used when hops are plentiful and never used to replace more than a third of the total hops. He also suggested that since they were all devoid of tannin, 1 lb of tannic acid should be added to every 100 lb of substitute.

CHAPTER 5

ADDITIVES AND PROCESSING AIDS IN THE TWENTIETH CENTURY

ANTIFOAMS

Antifoams are processing aids added to beer during fermentation in order to reduce the level of foaming. This allows more beer to be put into a fermenting vessel of a given size as the head takes up less space. It has the effect of increasing the fermenting capacity of a brewery without any capital expenditure.

Antifoams destroy the surface elasticity of the gas bubbles in a foam causing its collapse. Those which are used in beer must be non-toxic to micro-organisms as well as humans and must be insoluble so that they can be removed after fermentation. The antifoam commonly used in beer is dimethylpolysiloxane. It is used at the rate of around 4 to 8 ppm, which will reduce the normal head size by about 25%. Yeast adsorbs some of the dimethylpolysiloxane and this is removed by the normal filtration during processing. The rest is removed by adsorbants used at filtration. Although it is a processing aid and not an additive, small amounts may be left in the beer. These were considered to be so small that the FACC approved its use in 1978. In the U.S.A. the maximum residual amount which is allowed in beer is 10 ppm.

Various surface active lipids (e.g. glyceryl mono-oleate), other unsaturated monoglycerides and sorbitan monolaurate have also been tried as antifoams but are not in common use.

ANTIOXIDANTS

Antioxidants are used to reduce the levels of oxygen in beer when air becomes dissolved or entrained in it. The presence of oxygen in beer not only aids the growth of any spoilage micro-organisms present but oxidises certain substances to produce off-flavours. Protein-tannin complexes which produce chill haze are also oxidised to produce permanent hazes.

The two antioxidants used in beer are sulphur dioxide and L ascorbic acid. Sulphur dioxide is added in the form of sulphites which are permitted preservatives in beer in the U.K. Ascorbic acid, better known as vitamin C, is a basic dietary requirement so it is an acceptable additive.

Ascorbic acid is more often used in lagers. Care is needed in use as in the presence of iron it becomes an oxygen carrier. However, in conjunction with bisulphites which are reducing agents this does not occur, so they are often used together. It is used at a rate between 2 and 5 gm/hl, but if over used it can produce sulphury off flavours when there are low levels of air to scavenge.

Sodium erythorbate, an isoascorbate, is another antioxidant which is permitted for use in beer in the U.S.A. but is prohibited in the U.K.

BEER STABILISERS

These are additives which are used to reduce the amount of protein-tannin material in beer so that it does not appear as a haze in the finished product. This haze-forming material appears when the temperature of the beer falls, so is known as a chill haze, but in time and under certain conditions it becomes a permanent haze, when it is present at all temperatures. Beer becomes stable if either the proteins, the tannins or both are removed. The traditional method is to precipitate these compounds by prolonged cold storage, which is expensive. Additives offer alternative and cheaper methods of removal.

The additives can be separated into three groups:

1. Those which break down the proteins into smaller soluble compounds.
2. Those which react with and precipitate the proteins.
3. Those which adsorb either the proteins or tannins so that they can be removed.

PROTEIN BREAKDOWN AND PRECIPITATION

The additives used for breaking down proteins are enzymes and those used are discussed in this Chapter under Enzymes.

The substances used to react with the proteins and precipitate them are tannic acid preparations. Tannic acid is considered a safe substance and the amounts used are said to be less than those found in tea. The preparations are natural substances prepared from fractionated gallotannic acid. Early preparations acted slowly and were added to the storage tanks where the proteins precipitated. The modern preparations act quickly and are added in line before rough filtration stage and after cold storage. The precipitate is removed in the filter.

ADSORBTION

Adsorptive materials are used to remove either proteins, tannins or both. Proteins are removed either with bentonite or silica gel and the tannins are removed with polyvinylpolypyrrolidone (PVPP).

Bentonite is an insoluble alumino silicate, a naturally occurring type of clay. It adsorbs proteins when added to the cold beer tank to form a loose sediment which is removed. Its use has largely become outmoded as it also removes some head forming materials.

Silica gels are pure forms of amorphous silica prepared by acidification of sodium silicate solution. This forms a hydrosol which sets as a hydrogel and is then milled to produce a powder. In the hydrogel exceedingly small silica particles are held together to form a sponge-like structure. The spaces between the particles are filled with water to form a porous network. Careful control during manufacture produces a hydrogel which has a specific pore size and surface area. It is suitable for adsorbing haze-forming beer proteins, leaving the head-forming proteins and amino acids unaffected. The rates of use are between 30 mg/hl and 200 gm/hl. Silica gel is added in the cold room tank or just prior to filtration, which removes it and the proteins from the beer. It is a very inert substance which as it is removed from the beer presents no danger to the consumer.

In the late 1950s research showed that polyphenols could be adsorbed onto polyamide resins. Nylon 66 was initially used and removed between 15% and 30% of the polyphenols when added at the rate of 0.5 lb/brl. However, it also adsorbed compounds which affected head retention and bitterness. The best results were obtained by adding it to the cold tank followed by filtration.

PVPP

Polyvinylpolypyrrolidone, usually referred to as PVPP, is an insoluble cross linked homopolymer of N-vinyl-2-pyrrolidone and was introduced as a beer stabiliser in 1961. It was preceded by the soluble polymer PVP which had been introduced in 1954. It prevents the formation of protein-polyphenol hazes in beer by adsorbing the polyphenols.

It can be used in various ways. The way in which it was first used was to mix it with the beer and to leave the two in contact for 24 hours or sometimes less. The solids were then filtered out using a powder filter.

The rate of usage was between 4 and 20 gm/hl of beer. Some more modern systems allow it to be recovered from the filter medium. A development often used is to mix it with silica hydrogels to increase its effectiveness. More recently it has been incorporated into filter sheets.

Care is needed in filter operation in order to prevent PVPP getting into the beer. Modern filtration plant in which PVPP is used minimises this risk and a trial conducted at a British brewery found complete absence of PVPP downstream from the filter. However, it was accepted in the Food Additives and Contaminants Committee Report on Additives in Beer 1978, that some must get into beer and be consumed, but they considered that the amounts would be so small that there would be no health risk. Some breweries fit polypropylene cloth guard filters to remove any particles which may escape the main filter.

PVPP is not considered to be a beer additive as it is removed from the beer, but there is usually a small amount of soluble PVP which is incorporated in the PVPP. This dissolves in the beer. In the Food Additives and Contaminants Committee Report on Additives in Beer 1978, they recommended a maximum level of 50 ppm for the level of dissolved material. The current regulations in the U.S.A. are that beer should not contain more than 10 ppm PVP and in Canada the level is set lower at 2 ppm. France does not allow the use of PVPP which it holds to be carcinogenic. It was a bone of contention between the French brewers and the Germans, that German beers which are allowed to be treated with PVPP were being imported into France, whilst their beers were being banned in Germany under their beer purity laws.

PVPP used in trials under extreme conditions, at higher concentrations and for a longer time than normal, were found to give the beer a flavour described as 'less than pure'. The cause was not identified.

COPPER FININGS

During the boiling of wort in the copper, insoluble material known as trub or break comes out of solution and is later removed. Copper finings are substances added in order to encourage its precipitation. The traditional ones used in the U.K. are red seaweeds (*Rhodophyceae*) and their extracts. *Chondus crispus* or *Gigartina* species are the ones which are the most efficient. The whole seaweed, which is the form most commonly used, is known as either Irish moss, carragheen or carrageen. Carragheenan or carrageenan is the preparation which is a purified extract of the seaweed made with alkaline hot water. It consists of a mixture of the ammonium, potassium, sodium, calcium and magnesium salts of the acid sulphate of a galactose polymer. It is the sulphate groups which are responsible for the interaction with the proteins of the break and produce the fining action. There are three types of polymers: kappa, iota and lambda, which have increasing numbers of sulphate groups. The kappa type has the optimum number for fining in the copper.

Irish moss is used in very small amounts – around a fraction of an ounce per barrel, corresponding to 10 – 80 ppm of the wort. The rate of use depends on the gravity of the beer being brewed and it is added about 20 minutes before the end of the boil. It is considered a processing aid as it is removed with the break, though amounts around 1 ppm may remain in the wort. However, it is an approved food additive in the U.K. and when the Joint Expert Committee on Food Additives reviewed the ADI level in 1984, it was reduced from 75 mg/kg body weight to 'ADI not specified', which in effect means that it is not considered a health hazard. The carrageenan used in beer should not be confused with the type known as 'degraded' charrageenan which has been chemically broken down into molecules of smaller molecular weight. Its use is prohibited in food as it produces ulcers and tumours in laboratory animals.

Various other seaweed materials have been used as copper finings, including furcelleran, sodium alginate and agar agar. Furcelleran is, like carrageenan, a galactose polymer which is extracted from Danish agar. Other materials, including bentonite, silicates, tannic acid, kieselguhr, isinglass, gelatin and acacia gum have also been tried but have not come into use.

FININGS

ISINGLASS

Although many fining agents have been used over the centuries, the one that is used exclusively as the primary fining agent of beer in Britain is isinglass, biologically known as *ichthyocoll.* Its properties were discovered in ancient times in the eastern Mediterranean region and the Middle East. The Ancient Egyptians knew that when they stored wine in bladders from certain types of fish, the wine cleared very quickly. It is of interest that they also considered these bladders a great edible delicacy and aphrodisiac.

It is not known for certain when finings were first used in the U.K., but records in the early seventeenth century show that it was imported from 'Muscovy' (Russia) , with an annual consumption of twenty-five tons which does not indicate widespread usage. However, in 1740 it is recorded that the porter brewers of London had universally adopted it.

The swim bladder of the sturgeon, which was fished for its caviar in Russia's northern rivers was the major source. The Dutch had the trade with the Russians so all supplies passed through the Netherlands, from where isinglass obtained its etymological origins. Its name is a corruption of the Dutch word Hyzenblas (Hyzen – to hoist and Blas – a bladder).

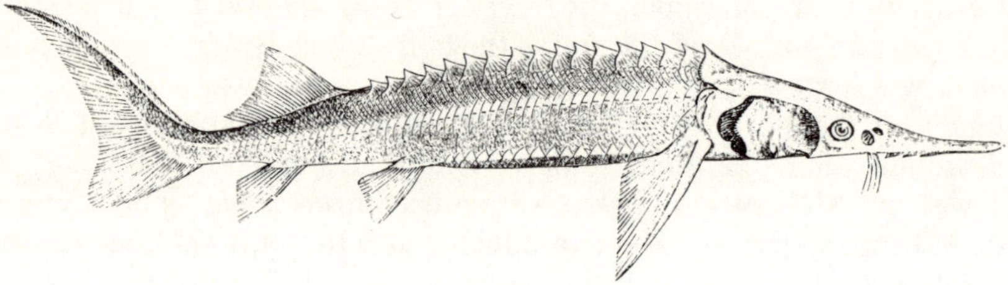

RUSSIAN STURGEON

According to Mathias, there were attempts to get supplies from a more reliable source in the latter half of the eighteenth century. A patent was taken out in 1760 by Humphry Jackson for making isinglass from fish caught in British waters. The blockade of the Baltic during the Napoleonic Wars quickly cut off supplies from Russia and precipitated developments. Brewers had to find an alternative and paid William Murdock £2,000 for the details of an isinglass substitute he had proposed to make in 1795. The process involved rubbing stale fish skins and tissue through sieves into a mixture of stale beer and vinegar. The business of manufacture was started in 1809 by James Butcher, but was soon stopped by the Excise as they considered the ingredients to be adulterants. However, in the ensuing court case the judge praised the accused for their efforts to find a substitute and production was allowed to continue. It was an inferior product but at least brewers were able to fine their beer. The substances which eventually proved the most satisfactory were cod and ling sounds and the skins of soles. Despite its inferiority, this type of finings continued to be used to some extent during the nineteenth century.

When the supply of Russian isinglass was resumed, most brewers went back to using it. However, it was facing competition from other countries who had found suitable fish in their own waters for isinglass production. The countries in competition were the Americas, India and China, but the Russian product has always been considered to be by far the best. Supplies are now mainly taken from the Far East, India and South America.

The names given to the various types sound very exotic but merely indicate the place of origin and the shape of the dried bladder. Long Saigon Leaf, which is obtained from fish of the threadfin family found in the waters around Saigon and Indonesia, was considered only second to the Russian isinglass when it was available. Other types from the thread-fin family of fish include Round Saigon, Penang Leaf and Penang Pipe. Some types also available are obtained from fish of the Catfish family and include Bombay Purse, Maracaibo Pipe and Brazilian lump.

Isinglass, Finings, and Brewers' Sundries.

BISULPHITED AND SOLUBLE CONCENTRATED ISINGLASS

Immediately convertible into tasteless Beer Finings of first purity and strength, at all times will act rapidly and permanently, a wonderful preservative and very economical. Bisulphited No. 1, 1/8; No. 2, 1/- per lb. Soluble No. 1, 1/4; No. 2, 0/10 per lb. In casks of 20, 40, 80, and 160 lbs.

LIQUID FININGS

Tasteless acidless finings, in kils., brls., or hhds., treble strength 1/2 per gallon; double strength 0/10 per gallon; brls. and hhds. treble strength carriage paid.

FININGS, 7d. pr. Gal.

A two-gallon cask of our Concentrated Finings, costing 20/-, will at once, by addition of water or mild ale, make into thirty-six gallons of strong finings. In 2, 4, 6, and 8 gallon casks. Casks free.

THE "PATENT" FININGS MAKING MACHINE.

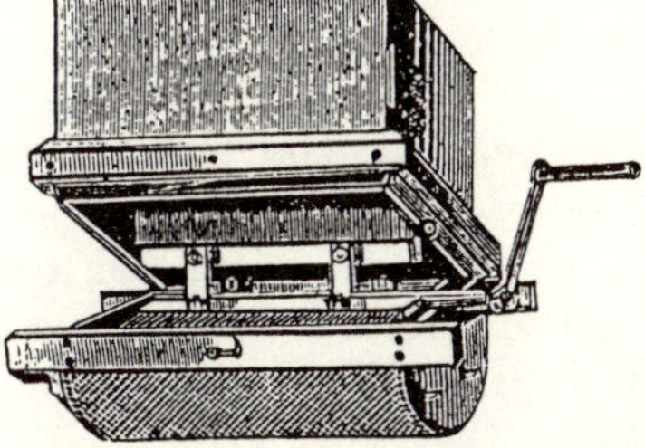

Over 2,500 now in use in the breweries of the United Kingdom and Colonies. In saving of labour alone quickly repays its cost. Finings made by it are every way superior to hand and sieve made, being perfectly aërated and of uniform consistency. The machine will sieve or strain thirty-six gallons in ten minutes. Numerous testimonials (received during past eight years) may be seen on application. Prices, to fit hhd. £5 5s. 0d.; brl. £4 14s. 6d.; and small or kil. size with one only sieve of intermediate perforation £3 10s. 0d. Larger sizes quoted for.

E. HERRING, GILES, & CO.,

WAREHOUSES AND STORES:

40, 41, and 42, GREAT TOWER STREET, and 23, BEER LANE, E.C.

CHEQUES CROSSED BANK OF ENGLAND. P.O.O. SEETHING LANE.

The swim bladder (or maw) is not found in all fish but it has several uses in those in which it is found. One of its uses is for adjusting the height of the fish in the water much in the same way as the air tanks in a submarine are used. When the fish are caught the swim bladder is immediately removed from the fish, opened and cleaned of undesired blood and tissue. It is then dried naturally in the sun before being exported. It is usually supplied as the raw leaf but in order to make it more presentable it is sometimes washed again, bleached with hydrogen peroxide, trimmed and flattened between rollers. A type known as strip isinglass also undergoes a further shredding process. Another type known as floc isinglass is produced by further maceration in a hammer mill. These additional processes make it easier to produce the finings.

The maw has to be liquefied in order to produce the finings which is done by 'cutting'. This involves soaking the maw in acid which slowly gelatinises it. In early times the acids used for 'cutting' were those produced in sour beer, but there was a high risk that microbial infection would be carried over to the finings. One solution to this problem was to sterilise the sour beer but the ultimate one was to use pure acids. Acetic, tartaric and sulphurous have all been used but the later two are the ones used today. Acetic acid is not used as it produces an unpleasant flavour. The cutting process takes any time between twenty-four hours and six weeks depending on how the maw has been prepared. The strip and floc preparations cut faster because of the increased surface area and time of cutting can be reduced further by high speed stirring.

Isinglass is almost pure collagen and one of the purest proteins found in nature. The molecules are a three strand structure twisted like a rope. They have a positive charge whilst yeast has a negative charge. The two attract each other and precipitate to form a compact sediment when finings are added to green beer. A rate of between 1 and 4 pints/barrel is used.

There have been several attempts over the years to make finings from bovine material. The middle layer of the skins of cattle is rich in collagen, but needs more treatment than that obtained from fish sources. This includes hydrolysis with sodium hydroxide. Recent developments in production of bovine finings have lead to the production of marketable products.

AUXILIARY FININGS

These are usually derived from one of three sources: red seaweeds, brown seaweeds and silicates. The red seaweed commonly known as Irish moss is the one most commonly used and is equally popular as a copper fining. Sodium alginate is extracted from brown seaweeds, usually *Laminaria*, for use as an auxiliary fining but it does not give as compact a sediment as Irish moss, so is less popular. Inorganic silicic acid preparations are the third type of auxiliary finings but tend to be unsuitable and return to silica gel. They also tend to produce a loose flocculum and can, at high rates of usage, give the beer a harsh flavour.

Auxiliary finings act in the same way on the proteins in beer as they do on the proteins in wort. They are negatively charged and flocculate together with the positively charged proteins. They also assist in making a compact finings sediment in the cask or tank as they also flocculate with the positively charged isinglass.

FOAM STABILISERS

Foam stabilisers are chemicals used by some brewers to give stability to the naturally occurring head which forms on beer when it is dispensed into a glass. The rate of collapse of the head depends on the amounts of head promoting substances in the beer and the amount grease or detergent on the glass which will destroy the head. In some parts of the country, such as the area served locally by Burton-upon-Trent, a head is not an important feature of a glass of beer, but in other areas a head is a very vital part of the drink and some brewers like to ensure that their beer always has a good head.

The use of foam stabilisers, or heading as it used to be called, can be traced back to the eighteenth century. The substances used at that time were ferrous sulphate and alum or ferrous sulphate by itself. The use of heading agents during this period is discussed in detail in Chapter 4.

Although the use of ferrous sulphate and alum was prohibited in the U.K. in the last century it is alleged that the use of ferrous sulphate continued until recent times in America, Belgium and France. One disadvantage of its use was that it formed a brown scum on the side of the glass. Jean de Clerk, the noted Belgium professor of brewing in his *Textbook of Brewing*, 1957, mentions the practice of putting a piece of iron in the conditioning tank in order to improve the head. He did not approve of the idea and pointed out the dangers of beer being contaminated with iron. Any use in Belgium must have been against enlightened scientific principles.

Cobalt salts were also known to produce very good heads on beer. In 1964 some brewers in the U.S.A. and Canada started using cobalt sulphate in their beers, followed two years later by some Belgium brewers. A number of drinkers began to develop symptoms which included the vomiting of blood and swelling of tissues. Eventually about fifty people died of heart failure because of heart muscle damage caused by the cobalt salts. Fortunately, the hospitals investigated the deaths of these drinkers, who had no previous history of heart disease, and connected it with cobalt salts in the beer which they had been drinking. Cobalt salts were immediately banned.

Though cobalt salts were the subject of a British patent they were not used by British brewers before the adverse effects were discovered. The two salts which were the subject of the patent were cobalt chloride and cobalt nitrate, the former being the most effective.

Research in 1964 showed that nickel salts were also effective but after the experience with cobalt salts their use was not pursued.

Metals are thought to act through a reaction with the isohumulones as they have no effect on foam when added to unhopped beer.

Other foam stabilisers used in recent times include natural gums, such as gum arabic, but it produced a loss in flavour because of poor efficiency and high levels of impurities. Methyl alginate, cellulose derivatives and a polysaccharide produced from *Erwinia* have also been advanced as alternatives.

The foam stabiliser which has gained most of the present market is propylene glycol ester (propane-1,2-diol alginate). It is an ester of alginic acid and considered a safe food additive in the U.S.A., the U.K. and other European countries. The Food Additives and Contaminants Committee in their Report on the Review of Additives and Processing Aids used in the Production of Beer, in 1978, were concerned that propylene oxide was used in its manufacture and that residues, which are undesirable, could be passed into the beer. They stated that a specification was necessary which would limit the contamination and recommended a usage level of not more than 100 mg/l. This exceeds the range of 40 to 80 mg/l at which the substance is effective.

It should be noted that foam stabilisers do not create the head on a glass of beer but stabilise the natural foaming properties. A small amount of grease will collapse the foam on an unstabilised beer in a matter of seconds, but foam stabilisers will prevent collapse. The small amount on a potato crisp can be used to show this and a packet of crisps can be used as simple detection kit for the amateur brewers' chemist. Heavier beers have more foam forming material in them which makes them naturally more stable without resorting to stabilisers, so stabilisers tend to be more used in lighter beers, particularly those which are brewed with a high percentage of adjuncts. They are widely used in the U.S.A.

ENZYMES

THE NATURE OF ENZYMES

Enzymes are found in, or are excreted by, living cells and their function is to increase the rate of reactions occurring in or around the cell at the pH and temperatures normally encountered. They are in effect biological catalysts and like all catalysts, although they take part in the reactions they are involved with, they are not altered by the reaction. As they are not consumed in the reaction they are only needed in very small amounts.

They are basically complex protein molecules with a three dimensional structure which makes them specific for a particular chemical reaction. The protein is linked to carbohydrate or lipid molecules in some enzymes.

Their chemical nature makes them very easily destroyed by heat, extremes of pH and inactivating substances. The inactivating substances such as heavy metals alter the active sites found on the molecule and prevent it functioning. There is an optimum temperature and pH at which enzymes work most efficiently, which is why temperature and pH have to be closely controlled during brewing.

The reactions which enzymes are capable of catalysing can be divided into those in which substances are synthesized and those in which substances are broken down into simpler substances. Both types occur during the brewing process but most are of the latter type.

The brewer carefully controls the conditions in which the enzymes function but sometimes he is limited because the enzymes in the malt are not present in sufficient amounts to do what he wants. At other times, in certain circumstances, he would like to use other enzymes which are not naturally present in malt. Commercially produced enzymes have been made available for the brewer to use if he has the need.

PRODUCTION OF ENZYMES

They are derived from either plant or microbial sources. When enzymes are produced from micro-organisms, the species selected is grown in deep culture fermentations in a factory. The cultures contain a complex liquid food supply which is kept under sterile conditions, with careful control of temperature, pH and oxygenation. When all the carbohydrate has been utilised by the micro-organism and the production of enzymes has reached its maximum, the culture is cooled. The micro-organisms are removed by filtration leaving the enzymes which have diffused from the cells into solution. The solution of enzymes may be concentrated to produce a liquid preparation or else the enzyme may be chemically precipitated so that a powdered preparation can be made.

THE USE OF COMMERCIALLY PRODUCED ENZYMES

In wort production, enzyme preparations may be used to replace the range of enzymes normally provided by the malt. This allows a grist to be used which can contain as little as 10% malt. The advantage to the brewer is that starch sources can be used which contain no enzymes and are cheaper than malt. Alternatively enzymes may be used to supplement those in the malt in both quantity and range if necessary. A common need for supplementing malt enzymes is when the malt contains a high level of β glucan, a substance which makes the wort very viscous and affects the quality of the beer. High levels occur naturally in barley in some seasons and cause brewers a great problem, but when an enzyme preparation of β glucanase is used to supplement the natural enzymes, the β glucans are easily broken down and the problem overcome. The main enzyme preparations used in the mash tun include: α amylase which partly breaks down starch to sugar, β glucanase as already mentioned and proteases which break down proteins.

In fermentations, enzyme preparations are mainly used to increase the amount of fermentable sugar available by breaking down the more complex unfermentable sugars into simpler sugars. Amyloglucosidase is the main enzyme used to break the necessary linkages in unfermentable dextrins to produce units of fermentable glucose. A highly fermentable wort is particularly desirable when a low residual sugar content is required in a beer, such as a diabetic or a dietetic beer. Beer used for shandy production needs as much of the sugar as possible to be converted to alcohol as it is later going to be mixed with sugar which is in the lemonade ingredients. It is therefore brewed with a wort containing the maximum amount of fermentable sugars.

After fermentation the same enzyme preparation can be added to a beer to produce fermentable sugars from unfermentable sugars in order to avoid having to use priming sugars. Amyloglucosidase is a little difficult to control for this purpose so a mixture of the two enzymes pullulanase and α amylase, which have the same total effect, have been used instead.

Proteolytic enzymes are added to beer after fermentation in order to modify the proteins which are potential producers of protein-tannin haze. They are added at the end of cold storage and include: papain from the latex of the papaya fruit (*Carica papaya*), bromelain from pineapple and pepsin from animal sources.

The enzymes in enzyme preparations occur naturally in the traditional ingredients of beer so it could be argued that they should not be regarded as foreign substances. The preparations may of course contain other substances which may or may not produce a risk to health.

Enzyme preparations are not controlled by any special regulations but they have to comply with the general provisions of the Food Act 1984. In 1989 the advisory committees of the MAFF were still in the process of reviewing their use. The situation before the review is that papain, pepsin and bromelain are classified in the group in which available evidence suggests that they are acceptable for use. The preparations of microbial origin which includes all the other enzymes mentioned are all classified in the group in which they are allowed for use, but about which more information is required for the current review.

HOP PRODUCTS

Hop products include all the modified forms of hops which are used instead of whole hops in brewing. None of them should contain any substance other than what is extracted from the hop. Out of the seven available types of hop product, the only two which are likely to contain contaminants are the two types of hop extract. The possible contaminants are solvents used to dissolve the extracts during their preparation.

Hop extracts have been in use since the middle of the nineteenth century. In 1863, Richard Morland patented a method of preparing a hop extract using steam. Another was patented in 1867 by Theophile Auguste Breithaupt again using steam; he had considered using alcohol but did not think it worth while. In 1869, John Henry Johnson on behalf of Hugh Burgess of Pennsylvania, U.S.A., patented a method of producing a hop extract using bisulphide of carbon either by itself or as part of a mixture with alcohol, ether or chloroform. Much of the development since then has been in the better use of solvents, not only to preserve the nature of the extracted substances, but also to improve the ease with which the solvents can be removed once the extraction has taken place. The solvents which initially became popular were low molecular weight alcohols, ketones, chlorinated hydrocarbons, hexane or various mixtures of these chemicals. For many years there was little awareness of the dangers of residual solvents, and in the U.S.A. the Food and Drug Administration (FDA) regulations, in 1969, allowed 2.2% residual solvent (methylene chloride or hexane) in kettle copper hops. In the 1970s there was growing concern over residual solvents in food, particularly the chlorinated ones. Chloroform was banned in the U.S.A. Although up to 2% of residual solvent was allowed in the extract it was demonstrated in Germany, in 1979, that after a good boil of thirty minutes in the copper the amount of solvent, which in the case under examination was methylene chloride, was reduced to 0.2%.

However, work had already begun in the early 1970s at the BRF to discover more acceptable solvents. They investigated the use of liquid carbon dioxide which had been looked at in Russia some ten years earlier. They found it to have several properties which not only made it acceptable from a food safety point of view but also made it very suitable for extracting the desired hop components. Carbon dioxide and ethanol both occur naturally in beer and are now used almost exclusively in the U.K. as hop solvents because of their acceptability. Whilst carbon dioxide was being developed as a solvent the technological advances in the use of other solvents had reduced their residual levels considerably. In 1983, levels of methylene chloride used in German extracts were quoted to be below 0.01% (w/w). There is still, however, some concern over the use of methylene chloride, as in the U.S.A. the National Institute of Environmental Health Sciences has conducted tests which show that it may be carcinogenic.

LIQUOR TREATMENT

Liquor is the name given in breweries to the water used for brewing. Certain mineral salts were identified in natural water supplies from various locations which produced superior beers of a particular type. Liquor treatment is the practice of adding or removing mineral salts, to give the brewing liquor properties similar to the most suitable natural supplies. At one time most breweries had their own supply which varied little in its mineral content and a liquor treatment could be formulated to suit the supply. At present, where a brewery is taking its supply from a water authority, water can be taken from several sources which are changed according to the demand of the customers and availability from the sources. Water supplied one day could be quite different in its mineral content the following day. Some brewers find the problem so great that they remove all salts from the water and add what they require rather than try to adjust the amounts which are naturally present from day to day.

Natural water supplies all contain some salts which are dissolved from the geological strata through which they have to pass. Perhaps the brewing liquor best known for the suitability of its natural mineral content is that of Burton-on-Trent, the town which became famous for its excellent pale ales. It was there that C. W. Vincent, in 1878, analysed the water to find out which salts gave it this special property. Through his work brewers in other parts of the country, with less suitable water supplies, were able to adjust the composition of their supplies to make them the same. The practice became known as Burtonisation. The water from other towns and cities which were famous for other types of beer were also analysed, enabling specifications to be made for liquors suitable for brewing all types of beer. Examples include:

Burton-on-Trent for pale ales.	Permanently hard water.
Dublin for stouts.	Temporary hard water.
London for brown beers.	Temporary hard water.
Munich for dark lagers.	Temporary hard water.
Pilsen for pale lagers.	Soft water.

When mineral salts dissolve in water they ionise. Gypsum, for example, which causes permanent hardness in water, breaks up into a positively charged calcium ion and a negatively charged sulphate ion. This is chemically expressed as:

$$CaSO_4 = Ca^{2+} + SO_4^{2+}$$

When water which contains a mixture of mineral salts is analysed, the analysis shows the concentrations of the individual ions in the water and not the concentrations of the individual salts; although the information is relevant, as it is the individual ions which characterise the brewing liquor as being suitable or unsuitable. However, for academic interest, with some knowledge of the strata from which the water was obtained, an intelligent guess can be made as to the combination of salts which produced the ions. The properties of these ions have to be looked at in order to understand why certain minerals are added to beer and others are not.

The principal ions present in most brewing liquors are calcium, magnesium, sodium, potassium, sulphate, chloride, bicarbonate, carbonate and nitrate. Other elements are present but only in small amounts and are known as trace elements. Some are essential but are not usually added as a part of liquor treatment. It should be mentioned that brewing materials are also a source of some of these ions; most of the magnesium, potassium and phosphate in wort come from this source.

CALCIUM IONS

These are very important as they affect the acidity of the wort. They react with phosphates in the malt and can form primary, secondary or tertiary phosphates. The secondary phosphate is sparingly soluble and the tertiary phosphate is insoluble. The more calcium present in the wort the more tertiary phosphate there is produced, which is accompanied by a release of hydrogen ions making the wort more acidic.

$$3Ca^{2+} + 2HPO_4^{2+} \rightarrow 2H^+ + Ca_3(PO_4)_2$$

BICARBONATE AND CARBONATE IONS

These are undesirable as they take up hydrogen ions from the wort to produce water and carbon dioxide. The decrease in acidity which this causes is undesirable. When calcium bicarbonate is present in water producing temporary hardness the detrimental effects of the bicarbonate ions outweigh the beneficial effects of the calcium ions.

$$CaCO_3 \overset{Ca^{2+}}{\rightleftharpoons} CO_3^{2-} \overset{H^+}{\rightleftharpoons} HCO_3^- \overset{H^+}{\rightleftharpoons} H_2O + CO_2$$

MAGNESIUM IONS

These have similar properties to calcium ions and also react with phosphates in the malt, but the phosphates produced are more soluble, so fewer hydrogen ions are released and the increase in acidity is therefore less. If they are present with calcium ions they interfere with their action during mashing. They also impart a sour or bitter astringency to the beer in concentrations over 15 mg/l, but their presence is required in very small amounts as an enzyme cofactor.

SODIUM IONS

At low concentrations they influence sweetness but give a salty taste when a certain concentration is reached which turns to a harsh and sour taste when they are present in excess. When sodium sulphate is added to brewing liquor the combined effect of the ions on the beer is unpleasant but when sodium chloride is added at the correct concentration palate fullness is produced.

POTASSIUM IONS

These have similar properties to sodium ions but do not give beer the sourness that sodium ions do. If chloride ions need adding in a liquor treatment potassium chloride is therefore often used in preference to sodium chloride. Above 10 mg/l potassium inhibits certain yeast enzymes and makes beer laxative.

SULPHATE IONS

These are said to give a dry flavour to beer and increase the perceived bitterness. This is appreciated in bitters and pale ales.

CHLORIDE IONS

These give sweetness and fullness to beers. In taste tests sodium chloride added to beer tasted no more salty than beers to which calcium chloride was added. Pronounced saltiness was only detected when the level of sodium chloride was added so that the level of sodium was in excess of 150 to 200 ppm.

MAGNESIUM IONS

These give a bitter/sour flavour but it depends on the balance between calcium and magnesium ions.

FERROUS IONS

These give a metallic taste to beer at low levels between 0.5 and 1.0 ppm. They also causes haziness, rusty deposits and slime but are essential in trace amounts.

Salts containing some of the above ions are included in liquor treatment but other chemicals are also added for the following:

ADJUSTMENT OF pH WITH ACID

The addition of gypsum to the mash is the traditional way of reducing pH in the U.K., but adjustment can also be made by adding acids. Sulphuric and/or lactic acids are usually used as they give a better flavour, but phosphoric and hydrochloric have also been used. This method of pH control is more popular in lager brewing than ale brewing. The Germans, who are not allowed to use acids, sometimes allow a little natural lactification of the mash with lactic acid bacteria to achieve the same effect. A low pH is beneficial to the beer as it improves break or trub formation, stabilises alpha amylase, reduces hop utilisation, reduces colour formation, precipitates undesirable oxalates, helps yeast flocculation and improves foaming by reacting with lipids.

SOFTENING OF WATER

The presence of bicarbonate ions are responsible for temporary hardness and there are a number of ways in which they are removed:

1. By boiling which precipitates insoluble calcium carbonate.

$$Ca(HCO_3)_2 = CaCO_3 \downarrow + H_2O$$

2. By means of ion exchange resins.

$$RH + Ca(HCO_3)_2 = RCa + 2H_2O + 2CO_2 \uparrow$$

The resin remains in the softening apparatus

3. By the addition of measured quantities of slaked lime.

$$Ca(OH)_2 + Ca(HCO_3)_2 = 2CaCO_3 \downarrow + 2H_2O$$

The calcium carbonate precipitates.

4. By the addition of acids. Sulphuric, lactic and phosphoric are used. Sulphuric is preferred as it does not produce salts which affect pH, but others are used where sulphates are already high.

$$H_2SO_4 + Ca(HCO_3)_2 = CaSO_4 + 2H_2O + 2CO_2 \uparrow$$

Softening of permanently hard water is sometimes carried out if there is too much gypsum in it, as it can give the beer an undesirable softness. This is carried out by adding sodium carbonate but is a rare form of liquor treatment.

CHLORINATION

This is carried out by all water authorities and sometimes by brewers in order to improve the microbiological stability of their water. Chlorine or sodium hypochlorite is used and the object is to produce a free chlorine reserve of 0.3 to 0.5 mg/l. It is removed in some breweries before brewing, by passing the liquor through activated charcoal or rousing it with potassium metabisulphite. Many breweries do not remove it and excess chlorination has sometimes produced chlorophenols which have a medicinal flavour. They have a very low taste threshold value in the region of a few parts per billion. Fortunately, most chlorine is driven off during boiling before it can produce any flavour in the beer.

SUMMARY

In order to indicate which chemicals are most commonly used in liquor treatment the following summary is given:

1. Most brewers dislike temporary hard waters and bicarbonate ions are removed by one of the ways mentioned. This is theoretically not necessary for dark beers but as fewer of these are being produced, unless a brewery is producing a lot of dark beers, it is easier to soften all the brewing liquor.

2. Permanently hard waters are liked, as desirable calcium ions which help reduce pH are present without the unacceptable bicarbonates of temporary hard waters. Gypsum is therefore added to calcium deficient worts.
3. Chlorides are desirable in fuller dark beers and are added as sodium or potassium chloride, though calcium chloride is also used.
4. Usually there are enough sulphate ions present because of the addition of gypsum. Sodium sulphate is rarely added as it can produce unacceptable flavours.
5. Magnesium ions are only occasionally added as magnesium sulphate.

The following chemicals are those which were recommended for use in brewing liquors and malt steep liquors by the Food Additive and Contaminants Committee in their report, in 1978.

Acetic acid
Calcium chloride
Calcium hydroxide
Calcium sulphate
Hydrochloric acid
Lactic acid
Magnesium sulphate
Phosphoric acid
Potassium bisulphate
Potassium chloride
Sodium hydroxide
Sodium sulphate
Sulphuric acid

MALTING ADDITIVES

Additives used in malting can be classified as follows:

1. Pesticides used during storage.
2. Chemicals used to adjust the composition of the steep liquor.
3. Additives used to reduce malting loss.
4. Stimulants for germination.
5. Additives for breaking dormancy.

PESTICIDES

Barley has to be stored after harvest until it is required for malting. During this period barley can be subjected to the attentions of various insects and mites. These include the saw tooth grain beetle, the grain weevil, the rust-red or flat grain beetle and various moths and mites. Insecticides and acaricides are available which are cleared for use under the Pesticides Safety Precautions Scheme and can be sprayed on grain as a preventative measure (e.g., malathion, pyrethins or γ-benzene hexachloride). When an infestation occurs, insecticide spray or fumigation is used. Hydrocyanic acid gas, methyl bromide and ethylene dibromide are examples of the substances used. Fungi are less of a problem during storage as they can be more easily controlled by the use of the correct temperature and moisture.

The FACC Report on the Review of Additives and Processing Aids in the Production of Beer, in 1978, saw no problems with pesticides which had been cleared under the Pesticides Safety Precautions Scheme. There has been some public concern over residual pesticides on food, but this was initially over those left on fruit, nuts and other food which is eaten unwashed. In the malting process the grain is steeped in water for several hours, with a number of changes, which does much to remove any pesticides on the grain.

It is interesting to read the following account of the problems weevils caused for brewers in the eighteenth century and to consider which situation is preferable.

> [Concerning a small town twenty five miles from London.] Now according to the opinion of some, all or most of their malt lofts or granaries in the same are infected by the weevil, and have been so for many years past, insomuch that they despair of ever getting rid of their constant lodger. If then this horrid creature breeds in such abundance in one small town, and accompanies the malt alive both before and after grinding, even into the mash tun, the drinkers in course must swallow the quintessence of their nasty bodies; which before, on sight of only one being crushed under the nail, would be apt to turn his stomach. And therefore it highly concerns all to be assured their beers and ales are brewed from sound malt, clear of that stinking insect, the weevil.

STEEP LIQUOR

Soft water, which is slightly acidic, is adequate for steeping, but alkaline waters are preferred, so chemicals are sometimes added to achieve this. A weak solution of 0.05% to 1.0% sodium carbonate or sodium hydroxide is usually used, though calcium hydroxide which is less effective can be used. They help remove the testinic acid, which is a mixture of acidic, proteinaceous and phenolic material leached from the grain. Calcium or sodium hypochlorite is sometimes used to prevent bacterial growth in the steep liquor but there is a danger that medicinal flavours will be produced in the beer.

MALTING LOSS

Malting loss is loss in the dry weight of the grain from when it is barley to when it is malt. Apart from the loss which occurs when substances in the grain pass into solution during steeping and when the rootlets on the germinating barley are removed, there is also a loss of carbohydrates through respiration. Respiration provides the energy for the grain to start growing and comes from the breakdown of carbohydrates to carbon dioxide and water. Sodium and potassium bromates are used to check the amount of respiration during malting. It stops excessive root growth which is removed from the final product and cuts down the amount of heat produced, which is also desirable. However, the chemical changes known as modification still occur in the grain.

The use of these chemicals has not been popular, although they are also used in baking as flour improvers. Most people associate bromates with preparations used to cool sexual drive. It was anticipated that they would be recommended for prohibition by the FACC in their report on beer additives, in 1978, but they were not. Their presence in beer and malt has been investigated and it has been found that they do not survive the boil in the copper, so do not get into the final product.

The FACC was also asked in 1978, when they made their report, to consider the use of the following additives for the same purpose: sulphur dioxide, sulphuric acid, phosphoric acid, hydrochloric acid, citric acid, acetic acid and lactic acid. They recommended that their use should continue although they are not in widespread use.

It is necessary when germinating grain is put into the kiln that respiration is stopped straight away in order to cut malting losses. It takes some time before the grain becomes warm enough for the growth-stimulating enzymes to become inactivated. Sulphur dioxide in the kiln gases is used as an additive to achieve quicker inactivation. It can, however, cause a condition known as magpie malt. Sulphur trioxide can form which burns black patches on the surface of the malt. Ammonia is a permitted additive which is used to prevent this.

GERMINATION AIDS

With the onset of germination during malting, the endosperm starts being broken down by enzymes. The production and release of these into the endosperm is stimulated by gibberellins which are plant hormones. Gibberellic acid (GA3) is added by some maltsters, if their customers permit its use, in order to supplement the natural gibberellins and stimulate the germination of the malt. As it can reduce germination time by 20%, it can save considerable money in the cost of producing malt. Gibberellic acid (GA3) occurs naturally in barley, but it is produced commercially from culture filtrates of the fungus *Gibberella fujikuroi*. The FACC in its report on additives in beer, in 1978, recommended its use be continued, but the Toxicity Sub-committee recommended that the residues in beer should not exceed 0.05 mg/l. It is impossible to distinguish between natural gibberellins and those which are added, so it was recommended that the limits should apply to the total gibberellins.

DORMANCY

Dormancy is a natural condition in grain, whereby it will not germinate until some time after it has been harvested. It evolved so that the grain does not germinate whilst it is still attached to the parent plant. The length of dormancy varies depending on the weather, the conditions under which the grain has grown and the variety of the plant. Whatever the delay, it is sometimes a problem for the maltster. One method of breaking dormancy, though not widely used, is to add hydrogen peroxide to the steep liquor. This breaks down releasing oxygen which helps overcome the problem. Since hydrogen peroxide breaks down to produce oxygen and water there is no risk in its use.

PRESERVATIVES

The preservatives and the permitted amounts which can be used in beer in the U.K. are found in the Preservatives in Food Regulations.

PERMITTED PRESERVATIVES IN THE U.K.

There are seventeen permitted preservatives on the list which is given in full in Appendix 4. Despite the apparently large number, in practice the sulphites are the only ones which can be used. The permitted maximum amount allowed is 70 ppm sulphur dioxide which they all produce when dissolved in beer. It is not very practical to add sulphur dioxide itself to beer as it is a gas, so sulphites which are liquids or powders are preferred. There are three types of sulphites: the ordinary sulphite such as Na_2SO_3, the bisulphite (or hydrogen sulphite) such as $NaHSO_3$ and the metabisulphite such as $Na_2S_2O_5$. The metabisulphite is produced by removing a molecule of water from two molecules of bisulphite. Solid sodium bisulphite has never been isolated as when solutions produced during manufacture are evaporated to produce solids a molecule of water is also driven off to produce the metabisulphite. The metabisulphite salts are the ones most commonly used. On dissolution bisulphite ions, sulphite ions and sulphur dioxide are produced, the proportions of which depend on the pH.

The other permitted group of preservatives is benzoic acid and various benzoates. The maximum amount permitted in beer in is 70 ppm. They are not suitable for direct use in beer because of the risk of affecting the head, although they are used in soft drinks where this does not matter. They are included in the list of permitted preservatives as they are used as preservatives in some priming solutions and enzyme preparations which are added to beer. The minute amounts passed onto the beer from these sources does not affect the head.

Emulsions of the antifoam agent dimethylpolysiloxane are permitted to contain up to 1000 ppm sorbic acid and therefore the amount that would be passed on to beer when it is used at the correct dosage is also permitted.

OTHER PREPARATIONS

One of the benzoates not permitted for use in the U.K. is n- heptyl p-hydroxybenzoate. It was the first anti-bacterial agent allowed for use in the brewing industry in the U.S.A. where it is fairly widely used and has FDA approval (U.S. 1965). It has two disadvantages in that it adversely affects head retention and it can come out of solution to form a haze under certain conditions because it has to be used near to its threshold of solubility. It is usually used with a head stabiliser such as propylene glycol alginate in order to overcome the first problem. It was considered for use in the U.K. but the FACC advised against its use in 1978, because of doubts to its technological need and toxicology, which had been expressed in an earlier report in 1972.

A related compound which is more soluble and has little effect on the head of beer is trihydroxybenzoate, also known as n-octyl gallate. This is also prohibited in the U.K. but its use was allowed in the U.S.A. where it was recognised as being generally safe. However, at the recommended rate of usage of 100 ppm, the recommended maximum daily intake of 0.2 mg/kg would be exceeded by a moderate drinker, so it is not used.

Another substance which was used in the U.S.A. and approved of by the FDA (U.S. 1968) was diethyl pyrocarbonate. The attraction is that it breaks down to ethanol and carbon dioxide which are naturally found in beer in large amounts. However, it was discovered by two Swedish scientists, in 1971, that despite its quick breakdown it still had time to combine with ammonia to produce urethran. This was discovered in 1943 to be carcinogenic. In view of the fact that some brewing materials and brewing liquors contain naturally occurring ammonia, it was decided that it was not safe and was banned in 1972. Its use was first investigated in the U.S.S.R. and then it was developed in Germany. It is expensive to use but despite this it has been used in the U.K. as a sterilant. The attraction is that it decomposes if it is not rinsed off the article being sterilised to leave supposedly harmless compounds. However, it has never been used as a preservative.

SULPHUR DIOXIDE

Sulphur dioxide is used in various parts of the malting and brewing process, and though it is not used as a preservative in all cases, only those that are will be considered at this point.

Sulphur dioxide used to be added to hops during drying in the oast. It was produced by burning sulphur in the kiln. This gave the hops a uniform bright yellow tinge which did not deteriorate as fast as the colour of unsulphured hops. It was also said to prevent the formation of raw flavours caused by deterioration of the resins. However, it was found from the analysis of essential oil from hops, that those hops which had been treated with elemental sulphur for hop diseases in the field or had been sulphured on the kiln, contained more sulphur compounds than those which had not. Some of the sulphur compounds from these hops to some extent survive boiling in the copper, so that amounts above their flavour threshold value remain. If the hops are used as dry hops this is even more likely to occur. The flavours of these compounds have been described as cheesy, cabbagey and like burnt rubber. In 1981 it was decided to ask the Hop Marketing Board to stop growers using sulphur on the kiln and in the field beyond the burr stage in cultivation.

Sulphites are sometimes added at the beginning of fermentation in order to try to kill off any wort-bacteria which may be present in the wort. Wort is particularly susceptible to the effects of spoilage at this stage as there is a lag phase of a few hours before the yeast starts reproducing. Some of the wort bacteria are more virulent than the yeast and can spoil the wort before the yeast reduces the pH of the wort, which stops their growth. Substances containing sulphites used for this purpose are described by the manufacturers as yeast purifiers.

Sulphites are most commonly used by brewers as a preservative which is added to beer after the end of fermentation in order to prevent the growth of bacteria which may spoil the beer. Sulphur dioxide also serves as an oxygen scavenger. Much of it reacts with oxygen, so levels found in finished beer are usually well below the permitted maximum.

Sulphur dioxide is a permitted preservative in the U.S.A. but not in Germany. However, the Germans are allowed to use it to treat hops, malt and water because most of the sulphur is driven off during boiling in the copper and the amount remaining is very small compared with the amounts produced during fermentation. Sulphur dioxide is produced during the most active part of the fermentation from sulphates in the brewing liquor. In some cases the amount produced during fermentation has exceeded the German legal maximum of 10 mg/l, though this has been mostly in stronger beers. Most other beers are well below this limit.

Sulphur dioxide in food can have quite adverse effects on asthmatics; if they are sensitive to it an asthma attack can result. Sulphur dioxide in beer is liberated in the mouth and inhaled when the beer is swallowed. The small amount is barely noticed by most people but because it is an immunological reaction in asthmatics only very small amounts are needed to produce an asthma attack.

High levels in beer can cause gastric irritation, nausea or diarrhoea. People with damaged livers or kidneys can also be affected. The Accepted Daily Intake (ADI) level is 0.7 mg/kg. This is about 49 mg for a 70 kg (11 stone) person. Bearing in mind that other foods also contain sulphur dioxide a person drinking beer with high levels could easily exceed the ADI level. As ingredients do not have to be declared on beer labels drinkers who are sensitive to sulphur dioxide are put at a disadvantage as they have no indication of how much they are consuming with their beer. They only know that they are consuming no more than 70 ppm. In the U.S.A. they are more sensitive to the problem and if levels above 10 ppm are used, the use of sulphur dioxide has to be indicated on any labels.

NISIN

Nisin is a polypeptide which is an effective bacteriocin used in certain foods but not beer. It is under consideration for an E number and is being investigated by some researchers for use in beer. Research has shown that it is particularly effective against lactic acid bacteria which are very difficult to eradicate from stocks of brewery pitching yeast. It also has the advantages of producing no effects on strains of brewing yeast and no deleterious effect on the flavour of beer.

The two potential uses in brewing are to add it to fermentations as a preservative measure or to add it as a remedial measure when lactic acid bacteria are detected. However, as the law stands at the time of publication it can only be added to foods in hermetically sealed containers, so it could only be added to bottled or canned beers. It has been suggested that initially it could be used to wash yeast, as residual amounts in beer after washing should be within permitted limits.

YEAST FOODS

Yeast has a fairly complex range of nutritional requirements; a balanced and well produced wort contains all of them. A characteristic of some yeasts is that over a period of time performance can fall off in some respect. This can become evident in a number of ways, such as failure to completely attenuate the wort, a reduction in the speed of fermentation, a failure to produce a good yeast head during fermentation and the production of off-flavours. This has often been found to be due to the absence of some nutritional requirement in the wort. Wort composition can be very difficult to control because brewing materials are products of nature, varying from season to season and from one source to another. If a brewery does not have extensive laboratory facilities with which the wort can be analysed to detect any shortages, it has often been the practice to add yeast food as a form of blanket protection. Yeast foods usually contain a cocktail of minerals, vitamins, amino acids and other nutrients which are likely to cover any deficiency and improve yeast performance. It is more economical for a small brewer to use yeast food than pay for extensive analytical tests.

There are several suppliers of yeast foods and they are somewhat less keen to reveal the composition of their product than they are to extol its virtues. This is understandable where there is competition and one's product is selling well. All that can be revealed are details of substances which are known to have been used in yeast foods, generally, over the years.

Amino acids are essential to yeast for protein production, so growth and therefore fermentation stop should any vital one in the wort become used up. They are therefore a common ingredient of yeast foods.

Ammonium salts are very simple nutrients which can provide nitrogen for yeast growth and are assimilated before the amino acids. They are therefore often included so that there is the least possible delay in the start of fermentation. This is an important factor in preventing spoilage of the wort by bacteria. Ammonium sulphate and ammonium chloride are two which are used.

Vitamins required by yeast are mostly those of the B complex variety and are obtained for use in yeast foods from extracts of autolysed yeast.

Zinc is toxic to yeast at high concentrations as it inhibits some enzymic action, but zinc chloride added at the rate of 0.1 to 0.2 mg/ml in yeast foods will stimulate fermentation. A wort can become deficient in zinc because of its affinity for insoluble material which is removed before fermentation. Zinc sulphate is another salt used in yeast foods.

Traces of iron and copper are also required by yeast and deficiencies in wort can occur though the situation is rare. Many brewers on the contrary reduce the amounts present, as they are potentially capable of causing haziness in beer because of their catalytic action. Copper sulphate, ferrous sulphate and ferrous ammonium sulphate can be added by the brewer if he feels it necessary, as they were approved for use by the FAC in their report on additives in beer, in 1978.

Manganese is another trace element added to yeast foods as it is required at levels below 0.2 ppm as an enzyme cofactor.

It is likely that Pasteur was the first person to realise the importance of yeast foods. One yeast food which was used at the end of the nineteenth century was known as Pasteur's Solution and consisted chiefly of a mixture of ammonium tartrate and calcium and potassium phosphates. Another yeast food which was patented by Rands, in 1878, was ammonium phosphate and used at the rate of 1 lb per 50 bushels of grain.

YEAST WASHING

In the U.K. one of the ways of dealing with the contaminant bacteria which grow in stocks of brewery pitching yeast is to wash it with suitable chemicals. The basis of the practice is to adjust the concentration of the chemicals so that the bacteria are killed whilst the more robust yeast survives.

Chemicals used include: bisulphite preparations, acids and acidified ammonium persuphate. Thc acids usually used are: tartaric, sulphuric or phosphoric, which are used to reduce the pH to between 2.2 and 2.4 for two hours. Ammonium persulphate at a concentration of 0.7% and acidified with sulphuric acid to give a pH of 2.8 for 16 hours has also been successfully used. After washing most of the substances used are rinsed out of the yeast with water before it is pitched into a brew. Yeast washing has become more widely used in recent times with concern over high ATNC levels caused by bacterial infection.

Antibiotics have been evaluated for yeast washing but are not used in the U.K. Nisin, a permitted food preservative, has recently been investigated for use in beer and one of the suggested applications has been for yeast washing.

CHAPTER 6

CONTAMINANTS FROM 1700

TRACE ELEMENTS

The amounts of many elements found in foods are often very small and measured in parts per million. These amounts in the past were below, or approaching, the limits of analytical detection so were known as trace elements. Certain trace elements are known to be essential for human life. In 1949 the list included six: copper, zinc, iron, manganese, iodine and cobalt. By 1977 the list had been extended to fourteen and the following were added: molybdenum, selenium, chromium, nickel, tin, silicon, fluorine and vanadium.

Other trace elements are known to be toxic at very low levels and no adverse effects are known of when they are absent from a diet. They include lead, arsenic, antimony, cadmium and mercury. They are not found naturally in food and only exist as contaminants unless they are deliberately being used to poison someone. Selenium is an unusual case in that it is an essential trace metal but is also toxic in larger amounts. The gap between the two amounts is very narrow. The same applies to fluorine. Other metallic elements are found in food in larger than trace amounts and include calcium, magnesium, sodium, potassium and iron.

The only two trace elements which have caused any significant problems in beer are lead and arsenic. Their effects, because of their toxicity, have perhaps been the most far reaching of any contaminant, adulterant or additive. As a result the levels permitted in beer are strictly controlled by law.

Whenever cases of poisoning by contaminants or adulterants are examined one feature is common in all cases: the variation in response between individuals. Some people are affected very quickly with severe symptoms whilst others are unaffected even though they have taken in similar amounts of the poison. This was noted in the eighteenth century when Sir George Baker was investigating Devon Colic. When he was asked to explain the phenomenon he stated simply: 'There is a remarkable difference between men.' It has since then been shown to be a statistical phenomenon known as normal distribution. Whenever characteristics are measured in a group of people, whether it be height, intelligence or resistance to toxic substances, if the different levels of the characteristic are plotted against the percentage of people exhibiting that level, a normal distribution curve will result. This is shown in the example below where the levels of lead at which people first exhibit symptoms of lead poisoning and the percentage of the people affected at each level are plotted against each other. As in all normal distributions the majority of people exhibit characteristics around an average whilst a few people are above or below the average.

This is of course borne in mind when legal limits are set in order that people with low levels of resistance are not affected.

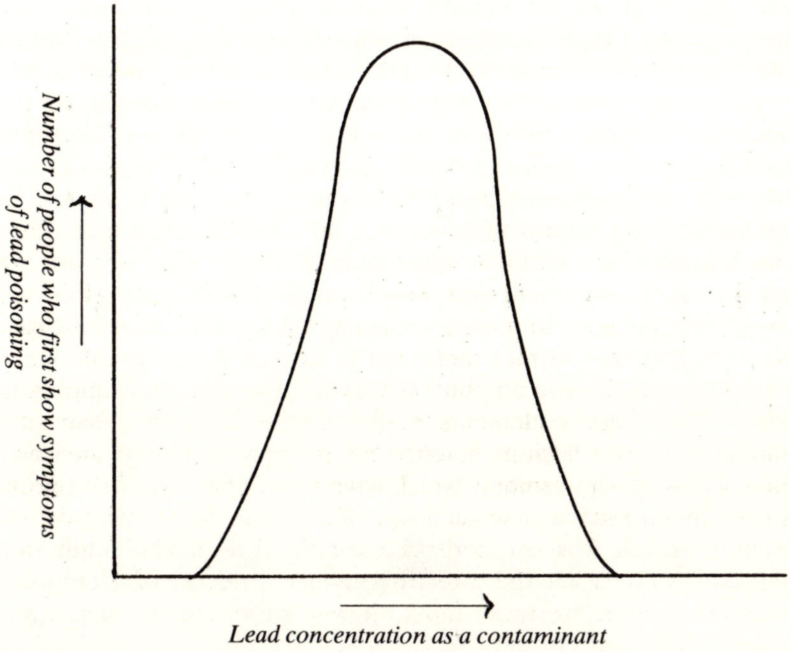

GRAPH SHOWING POPULATION SUSCEPTIBILITY TO LEAD POISONING

ARSENIC

Arsenic is a brownish-black or black metal which usually exists as the colourless tasteless oxide which was first produced by an Arab alchemist in the eighth century. It was a common poison with the symptoms of gastric pain, vomiting, diarrhoea, cramp, tingling in the feet, urine suppression, thirst, collapse and death. A dose of 100 to 300 mg is usually fatal. Many murders carried out with arsenic in the past were undetected because the symptoms are similar to those of cholera.

ARSENIC IN BEER

Beer has the distinction of being the means by which a large number of people were poisoned by arsenic at the end of the nineteenth century. This lead to the appointment of a Royal Commission under the chairmanship of Lord Kelvin in order that arsenic in food could be investigated. The report and the minutes of evidence were published in two volumes in 1903. The investigation was thorough and the evidence taken was very comprehensive, so much so that little more information on the subject has been gathered since that time. One person mentioned in the report had a tolerance to arsenic as high as 20 mg taken three times daily and another person on taking a single dose of 1.3 mg began to indicate symptoms of poisoning. The recommended limit of arsenic in liquid foods was set by the Commission at 0.14 ppm (0.01 grains/gall).

The arsenic which caused the outbreak of poisoning originated in iron pyrites used in the manufacture of sulphuric acid which in turn was used to manufacture glucose and invert sugar by a firm in Liverpool. These sugars were used in several breweries and beers were produced which contained between 5 and 15 ppm, with occasional samples containing 20 ppm. These beers affected 6,000 people in Lancashire and Staffordshire, 70 of whom died, excluding many more people who died whose cause of death was not correctly identified. There was also reason to believe that arsenical poisoning had occurred in Manchester before the main outbreak where it had been identified as alcoholic neuritis. Many gallons of beer had to be destroyed when the cause of the poisoning was discovered. Bostock, the firm who produced the poisonous sugar, was forced into closure.

Coal may also contain arsenical pyrites. Coke made from such coal has sometimes been used to make malt and if the vapours from the coke are used to directly heat the malt, it can become contaminated with arsenic. Several cases of poisoning occurred in Halifax in 1902, where beer made from such malt was found to contain 7 ppm.

BOSTOCK & CO.,

Works and Laboratory—

SOUTH DOCK, GARSTON.

ESTABLISHED 1860.

Manufacturers under Royal Letters Patent of the Finest Quality of

INVERTED CANE SUGAR.

STANDARD

MALT

Saccharine

SPECIALITY

CONE

SUGAR.

ALSO MANUFACTURERS OF

CARAMEL, or Sugar Colouring.

The whole of the Sugars turned out from our factory are exclusively manufactured for Brewers' use, and are all high-class qualities.

Further details and terms may be obtained at our General Offices, where we shal be pleased to show samples, or forward them on application.

GENERAL OFFICES AND WAREHOUSES—

30, JAMES STREET, LIVERPOOL.

BOSTOCK & CO.

LEVELS IN BEER

The recommended limits of the Royal Commission were, with the exception of levels in gelatin, not given any statutory force until 1959. However, they had been accepted as standards of good commercial practice and were recognised in the courts. When the Arsenic in Food Regulations did come into force in 1959 the legal limit for arsenic in beer was set at 0.1 mg/kg. Figures for the levels found in beer were published in the MAFF Food Surveillance Paper No.8 and showed that levels were less than 0.01 mg/kg. Total diet studies showed that the daily intake of arsenic in the U.K. was 0.1 mg per day. The maximum recommended daily intake set by the FAO/WHO in 1976 is 0.05 mg per kg body weight which is 3.5 mg for a 70 kg adult. The average person who drank a lot of beer would not be able to push his intake over the limits. Total diet studies incidentally showed that most of the dietary intake of arsenic was from fish.

LEAD

Lead has been used by man from the earliest of times; the Egyptians used it for glazing pottery and the Romans, who mined the ore in Britain, used it for making water pipes. It is a metal which is easily extracted from its ore and easily worked, so it has been used to fabricate thousands of different items used both domestically and industrially where its softness is not a disadvantage. It has, in more recent times, been used in batteries, solder, pigments and antiknock compounds in petrol. The world consumption in 1963 was approaching 2.5 million tonnes, in 1973 had passed 4 million tonnes and in 1983 had passed 4.7 million tonnes. Thus by now it is well distributed in the environment and taken into the body every day.

THE INTAKE OF LEAD IN THE DIET

The amounts consumed are generally very small, being measured in milligrams, so it is considered a trace element. Like the other toxic trace elements found in food it is generally present in amounts which cause no harm, even bearing in mind that lead also finds its way into the body in drinking water, other beverages and in the air we breath. The Food Additives and Contaminants Committee publication, *A Review of the Lead in Food Regulations, 1961* (HMSO 1975) put the average total intake of lead in food at 1 mg per week. A Department of the Environment survey, *Lead in Drinking Water* (1975-6) put the average amount taken in with drinking water and water based beverages at 0.2 mg per week. The amount taken in from the air in the worst case of 'non-occupational exposure' in an urban area was put at between 0.3 to 0.1 mg per week.

The values for intake with food appear to have decreased a little since the nineteen thirties when the earliest estimates were made. Monier-Williams, in 1938, put the lead intake from food at 0.2 to 0.25 mg/day whereas Kehoe in the same year had estimated it to be 0.16 to 0.28 mg/day. Figures taken from the results of total diet studies made since 1975 show that total daily intake is around the 0.1 mg/day level with indications of a slight decline. This gives a total level of less than a quarter of the 3 mg/week which is considered a tolerable maximum intake for adults by the Joint Food and Agricultural Organisation/World Health Organisation Expert Committee on Food Additives. It should be appreciated that these are only estimated average figures and that particularly with air and water, levels vary considerably in different situations. However, the figures do show how close we are to taking unacceptable levels of lead into our bodies.

Of the lead ingested 90% is passed out in the faeces and some in the urine and sweat. Some is deposited in soft tissue, hair and nails but most finds its way to the skeleton where it is stored. When present in a sufficient amount symptoms of lead poisoning begin to show. The early symptoms are constipation, muscular weakness and pallor of the skin. The most prominent and best known symptom is a painful colic. Eventually there is injury to the nervous system resulting in paralysis and tremors. The muscles on the back of the wrist are affected first causing wrist droop and later the muscles on the leg are affected causing a similar effect with the foot. Convulsions and death finally occur. Sight, kidney and blood problems are other symptoms.

When these unmistakable textbook symptoms occur it is known as frank lead poisoning. It occurs with different levels of lead for different people. Some people can have 150 μg/dl in their blood and exhibit no symptoms but other individuals can exhibit symptoms as low as 60 μg/dl. It is of interest to note that the average blood level concentration is about a quarter of that at which frank lead poisoning can occur. A WHO study group recommended an exposure limit of 40 μg/dl for men and for women over reproductive age. The level for women of reproductive age was set lower. The level of concentration in the blood in relation to uptake is dependent on a number of factors but statistically 100 μg/day (0.1 mg/day) from various sources will give a blood level between 20 and 40 μg/dl. Not all of our intake of lead into the body is absorbed and values published for the uptake of lead by the body vary between 1% and 35% depending on circumstances. An average value of 10% for adults is assumed by the DHSS.

LEAD POISONING

The occurrence of lead poioning from drink has an interesting history. Although he did not know what was causing it, the symptoms of lead poisoning from wine were first described, in 1617, by Francis Citois, physician to the King of France and the Cardinal Duc de Richelieu. An epidemic of colic had occurred amongst the inhabitants of Poitou and was attributed to the acidity of the wine. It was known as *Colica Pictonum*. The disease was also recognised in England, in 1703, by Dr William Musgrave who connected it with the drinking of cider. Further information on the disease was given, in 1739, by Dr Huxham in his publication *Opusculum de Morbo Colico Damnoniorum*. He recorded that there was a particularly severe epidemic in Devonshire which spread over the whole of the county. He also connected it with cider and noted that the drinking of cider had increased tenfold over the last thirty to forty years.

Sir George Baker investigated the occurrence of Devon Colic, as it had become known, and published his findings in 1797. He was a Fellow of the Royal College of Physicians and the Royal Society, and physician to Her Majesty's Household. He rejected the idea that the acidity of cider was the cause of Devon Colic after considering evidence such as the fact that other acid drinks did not cause colic, including cider made in Worcestershire and Herefordshire. Other evidence lead him to connect the colic with the presence of lead in cider.

The way that the cider had been contaminated with lead, was through the acid in the cider reacting with lead used in the cider making plant and equipment, or from lead or lead salts added to the cider to stop it going sour, particularly when unripe apples were used or if the cider was weak. The reason cider makers in Devon suffered from the problem, whereas the cider makers of Worcestershire and Herefordshire did not, was because the troughs in which they crushed their apples were made of several pieces of moor stones clamped together with iron hoops and sealed with molten lead. This reacted with the acid from the apples to produce lead salts in the cider. The Worcestershire and Herefordshire cider makers were not affected by this as putty was used instead of lead. Lead was also dissolved from pieces of lead sheet used to seal cracks on the press or even from lead sheet used to line the press.

Sir George Baker carried out several tests which conclusively proved the presence of lead in the affected cider. There was a vigorous outcry

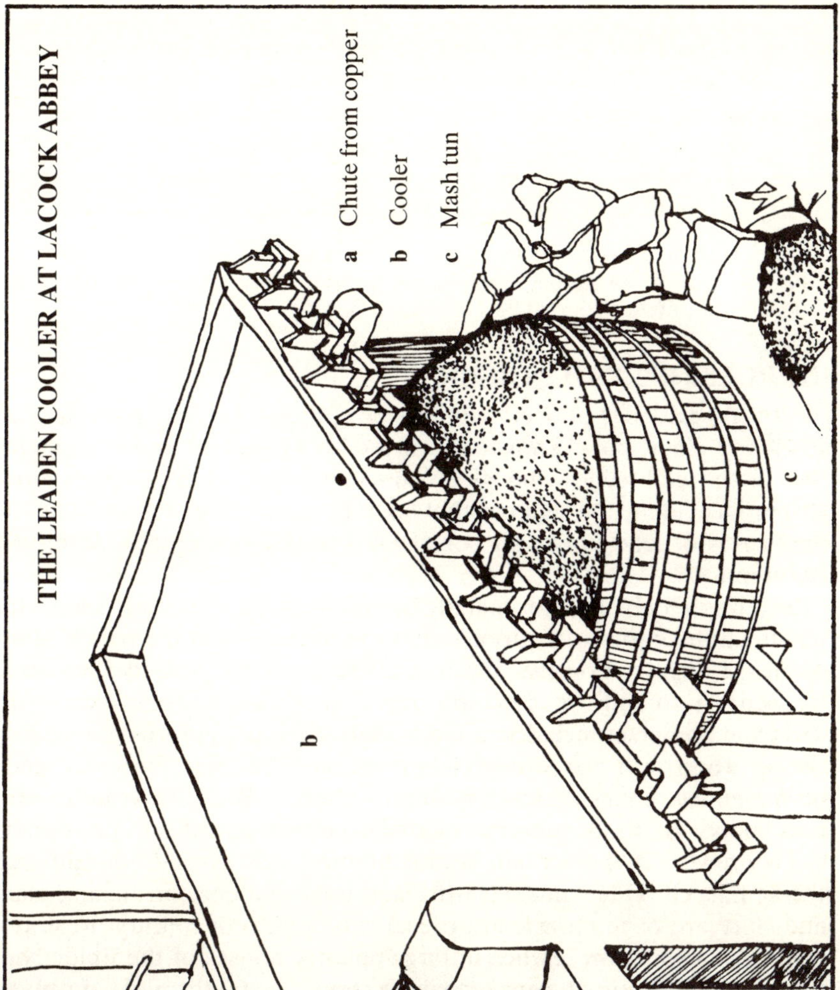

THE LEADEN COOLER AT LACOCK ABBEY

against his accusations by the cider makers, who employed several experts to reply to the accusations. Their answers were published along with Sir George Baker's essay, but the substance of the reply was so unreasoned and open to question that he had no problem in getting his explanation of the causes of Devon Colic generally accepted. At the beginning of his essay Sir George Baker made the following observation which sums up his approach to investigation and explains his success:

> Men partial to their own conceits now learned to attend to nature, to observe diseases and the genuine effects of medicines on the human body; and to make experiment the great basis of our reasoning.

SOURCES OF LEAD IN BEER

There is no evidence of beer ever causing epidemics of lead poisoning as did Devon cider, but a knowledge of the brewing plant and equipment used in the eighteenth and nineteenth centuries, offers enough circumstantial evidence to suppose that beer must have been responsible for some cases of lead poisoning, or at least some unacceptable levels of consumption.

Lead was widely used for making brewery vessels from early times. In 1335 in an inventory of Laurence de Long's brewery in the parish of St Martins, Ludgate, London, the first three items are: 2 leaden vessels, 1 leaden cistern and 1 leaden tap trough. Another brewery inventory of 1486 from Somerset includes: a tap trough of lead, 2 cisterns of lead for holding water and a steeping cistern of lead. Thomas Tyron in 1691 recommended lining coolers, mash tuns and fermenting vessels with lead, copper or tin despite the fact that he was a medical man rather than a brewer. If there is any doubt that this was carried out, anyone visiting Lacock Abbey in Wiltshire can see the evidence for themselves. In a courtyard of the house is a preserved eighteenth century brewery, which was a common feature of large country houses of the time. The open cooler which is situated below the copper is stoutly made of wood, but lined with lead. Southby, in 1885, records that fermenting vessels used to be lined with lead but that the practice had been discontinued, doubtlessly after the investigations of Sir George Baker had shown the dangers. During the present century beer has been kept in lead glazed tanks and was responsible for known cases of lead poisoning in 1922 and 1936.

Early hand pumps in small breweries were made with lead cylinders and wooden pistons and a surviving example of one was mentioned by Lloyd-Hind in his description of Queen's College Brewery, Oxford, in 1937.

The use of lead for water pipes in breweries was also common well into this century, until installations were prohibited in the 1930s. The larger water pipes in big breweries were usually cast iron. Beer dispense pipes in the public house were also of lead until after World War II when they were replaced by plastic. Lead pipes were recorded as being used in country districts as late as 1949 and isolated installations have been discovered in fairly recent times.

The danger of these lead pipes had been recognised for some time and a recommendation was made in 1929 that 15% tin should be incorporated into the lead in order to lessen the risk of contamination. It is doubtful if this was of any value. A more usual practice was to line the pipes with tin by 'tin washing'. Lead pipe is made by forcing lead near its melting point through a die to form a continuous pipe which is wound onto a drum. As it leaves the press molten tin from a ladle is poured down the open end of the pipe. Tin has a lower melting point than lead so stays liquid in the hot pipe running down it to leave a thin coating of tin. Unfortunately, in use the thin coating with its imperfections soon broke down exposing the lead. A more expensive pipe with a lining of tin 1 mm thick was far more effective. Lead pipes were known to be responsible for cases of lead poisoning. Two reported cases were men whose consumption was 1 gallon a day or more. The concentrations of the lead in the beer they drank were found to be up to 2.8 ppm.

This level of lead in beer was by no means uncommon as the following early data on the lead content of food shows (Food Surveillance Paper No. 10 MAFF).

	Range(mg/kg)	Mean(mg/kg)
1922–26	0.2 to 3.5	1.1
1927–39	0.2 to 21.5	1.8
1932–48	0.2 to 3.5	1.1
1938–42	0.1 to 1.4	0.47
1941–50	0.1 to 3.0	0.39
1937–44*	0.2 to 13.0	1.8
1937–44**	0.2 to 0.5.	0.12

* Country district ** Town district

It can be seen that the levels for country districts were much higher than the levels in town districts in 1937 to 1944. This was because there was greater use of lead pipes in country districts.

The larger diameter beer and wort mains in breweries at the beginning of the nineteenth century were generally either wood or lead. Charles Barclay, the well known brewer, giving evidence before a Parliamentary Committee investigating beer quality, in 1819, said that his beer mains were made of lead. After that time they were generally made of copper but were often tinned. Tinning of copper pipes involves coating them with an alloy of lead and tin and was used to prevent corrosion of the copper.

The problem of lead passing into beer from brewing plant and equipment was greatest when wort or beer was left in contact with it for a long period. Wort, for example, left overnight in a lead lined cooler or for a week in a lead lined fermenting vessel must have resulted in a high level of lead in the beer. The other factor which increases the pick up of lead is the presence of acids. Beer is not highly acidic but does contain a range of acids. Should it become more acidic through microbial spoilage when in contact with lead, its ability to pick it up is increased. Slowly souring beer left overnight in a lead public house cellar pipe can only have produce heavily contaminated beer. Though souring beer is not so common today it was more common in the eighteenth and nineteenth centuries when standards of quality control were lower. As mentioned in Chapter 3, sour beer was sold in certain areas at harvest time and elsewhere was blended with new beer to give it a taste of age.

With the disappearance of lead, brass and solder in brewing plant and equipment, the risk of lead pick up is now low. The last area of major concern was brass beer engines. Certain types of brass which were used in brewing plant and equipment can contain up to 2% lead. When EEC limits for lead in beer were lowered from a maximum of 0.5 ppm to 0.2 ppm, in 1978, there was a high chance that beer which had been in a beer engine with a brass cylinder, would pick up enough lead between serving periods to put it over the new limit. The brewing industry decided that before the Lead in Food (Amendment) Regulations 1979 came into effect, all beer engines would be modified so that they were made completely of non-lead-containing materials. Stainless steel cylinders and other parts of plastic were fitted instead.

LEAD IN WATER SUPPLIES

One of the other sources of lead contamination in beer is the water with which it is brewed. A survey of lead in drinking water, initiated by the Department of the Environment, was carried out in 1975 and the results published in 1977. It found that most of the U.K. water supply came below the limits set by the WHO. However, a good percentage of cases were reported where the maximum level had been exceeded. These mostly occurred where water had been standing in lead pipes for some hours. The houses where the ‘overnight first draw water’ was over the limit accounted for 20% of the total and those over the limit during the day accounted for 10% of the total. Since high lead levels in drinking water are being caused by pick up from lead pipes and are not present at source, it is unlikely that breweries who are not using lead pipes will be using water contaminated with high lead levels.

LEAD IN RAW MATERIALS

Barley is the material used in the greatest amount in beer, and the most likely cause of it having high lead levels is the pick up of lead from petrol-powered vehicle exhaust fumes. In order to establish the extent to which barley was picking up lead from this source, samples growing next to a motorway were tested. The average lead content of the samples growing on the east of the motorway was nine times higher than in the samples growing on the west. Samples contained between 0.004 mg/kg and 0.049 mg/kg, which was consider to be low. Most barley is grown away from traffic and in this situation it was found to contain no detectable levels of lead. Therefore, barley cannot be considered as a likely cause of high lead levels in beer.

LEAD LEVELS IN BEER

It was found, in 1982 , by Shaper et al. that in middle-aged men there was a connection between the lead levels in their blood and their alcohol intake. The COT advising the MAFF at that time, felt that the total exposure levels to lead were close to toxic levels for some people. In the light of this the Royal Commission on Environmental Pollution in its report, 'Lead in the Environment', recommended that more data be obtained on the levels of lead in beer and wine at the point of consumption. The results obtained were published in 1986. Draught, bottled and canned beers were all sampled. The 202 draught beers sampled were taken from licensed premises within 100 miles of Harwell, Oxfordshire. The contents of fifty bottles and fifty cans were also analysed. Two interesting points emerged from the results. The percentage of cans and bottles which had very low levels of lead (less than 10 μg/l) was about 90%, but in draught beer the level was much lower at 55%. Two of the draught beers had lead concentrations in excess of the legal limit of 200 μg/l.

In the discussion on the results, the authors calculated that the lead intake from beer by someone drinking 6 pints per day of a beer which contained 50 μg/ litre of lead would be 135 μg. The lead absorbed by the body from this is estimated at 20%, i.e. 27 μg of lead. This is high compared with the average amount which would be absorbed from food. They concluded therefore, that it was not surprising that some 'extreme' beer drinkers would have elevated lead levels in their blood. These figures also show how close some people are to frank lead poisoning. If the beer, instead of containing 50 μg/ litre, contained the maximum of 200 μg/litre, the uptake by the body would be 104 μg. As mentioned earlier, an uptake of 100 μg/day will give blood levels between 20 and 40 μg/dl and frank lead poisoning has been produced with levels as low as 60 μg/dl.

In 1981, a revised figure for the lead consumed with food, excluding beer, was put at 0.35 mg per week for men. In order to exceed the recommended maximum weekly intake of 3 mg per week, allowing for 0.1 mg being taken in from the atmosphere, a beer drinker would have to take in 2.55 mg with his beer. If the beer contained the maximum permitted level of 0.2 mg/litre he would have to consume over 13 litres, or just over 28 pints per week, or 4 pints per day. An estimate of the number of people drinking 6 pints per day (2.7 litres) was put at 500,000. If 90% were drinking draught beer and 1% of this was at or over the legal level for lead there are more than just a few people consuming high levels of lead.

FLUORIDES

Fluorine is the thirteenth most abundant element on the surface of the earth, but as it is fairly active it is not found in a free state. It exists mainly as minerals and was only produced on a large scale after World War II to manufacture fluorocarbons used as propellents and refrigerants. Most fluorine enters the body as fluorides in our diet. The total dietary intake in 1984 was estimated at 1.8 mg per day. Water which has not been fluoridated contains 0.1 to 2.0 mg/l so could raise the amount consumed by 2.0 mg, assuming an average consumption of 1 litre/day. Tea is one of the richest sources and has been known to contribute intakes as high as 5 to 10 mg/day. Toothpaste can contribute 0.5 mg/day.

High levels of fluorides in brewing liquor can affect the brewing process slightly and are passed through to the final beer where it becomes concentrated. High levels in beer are generally a result of high levels in the brewing liquor. Drinking large amounts of beer as with other minerals or contaminants increases the intake. Fluoridation of water supplies by local water authorities will result in even higher levels of intake.

Fluorine is necessary in the diet as it gives strength and stability to teeth and bones; its role in preventing tooth decay is well known. Excess intake causes fluorosis, the early symptoms of which are mottling of the teeth. Further symptoms are deposition of calcium in normally supple ligaments, tendons and muscles. The recommended dietary intake is 1 mg/day.

NITRATES, NITRITES AND N-NITROSO COMPOUNDS

Nitrates, nitrites and N-nitroso compounds are considered together as there is a link between the three.

NITRATES

Nitrates are very common in soil where they are produced by bacteria which fix nitrogen from the air. The use of man-made fertilisers has vastly increased the natural level in the soil and therefore the amount which is leached out into other strata has also increased. The concern over the increased levels in ground water and surface water, which pick up nitrates from these strata, has lead to national and international limits being enforced.

The main concern over increased nitrate levels in food is because they are converted to nitrites in the body by bacteria in the gastro-enteric tract. The nitrites cause methaemoglobinaemia which make the blood less efficient in transporting oxygen. This is of no detriment to most adults, but can cause concern in infants under three months old who are bottle fed with powdered milk made up with affected water. In adults the main concern is that the nitrites will react with amino componds such as secondary and tertiary amines to produce N-nitroso compounds, a process known as nitrosation.

N-nitroso compounds have been shown to produce cancers in laboratory animals, but although there is no evidence to show that they cause cancers in humans, the Brewing Industry and the Government are not prepared to take the risk that at some time in the future, some evidence will emerge showing that there is some risk to humans. They both welcomed moves to reducc thc levels in beer and other foods. The only loose connection with cancer that has so far emerged is that snuff dipping (as opposed to the more common practice of sniffing) has been linked to oral cancer. The only carcinogens that are known of in snuff are nitrosamines.

The main source of nitrate in the diet of humans is from vegetables. The amounts in different vegetable species varies and is dependent on the amount of fertiliser which has been used in their cultivation. Vegetarians will consume more nitrates in their diets than those who are not. The next largest source of dietary nitrate is water and it depends on where one lives on how much one consumes.

Total diet studies made in 1979, in order to estimate the total average dietary intake of nitrate, found the amount to be 61 mg/person/day. This figure ties in with studies made in other countries in Europe and in the U.S.A. which showed total dietary intake to be between 50 and 90 mg/person/day. The U.K. survey only attributed 8 mg of the intake to water but a survey of water consumption taken in 1980 indicated that the amount was higher than this. The mean consumption of tap water and tap water based drinks was shown to be 0.955 litres/person/day, varying between 0.39 and 1.57 litres/person/day. Taking Department of the Environment figures for 1985, which showed that the apparent median nitrate concentration of public water supplies was between 10 and 19.99 mg/l, the average intake of nitrate from water can be taken as between 10 and 20 mg/person/day.

The levels of nitrate in drinking water have been controlled mainly because of the concern over the effects of methaemoglobinaemia on infants. In 1970 the WHO European Drinking Water Standards set a desirable limit of 50 mg/l with levels up to 100 mg/l being acceptable subject to certain conditions. An EEC Directive was then brought into force in 1985 allowing a permitted maximum of 50 mg/l and a guide level of 25 mg/l with derogations permitted providing that there was no health hazard. In the U.K. derogations are permitted up to a maximum of 100 mg/l with a three month average no higher than 80 mg/l. In 1985 only 1.12% of the population was supplied with water containing more than 50 mg/l and none with levels above 90 mg/l.

The relevance of this to the beer drinker is that his beer is being made with this water and increased levels will generally result in increased levels in the beer. However, not all nitrate in beer comes from the water. A small amount also comes from the grist and the hops. The amount contributed by malt is estimated to be between 2 to 3 ppm and the amount from adjuncts to be less than 1 ppm. The amount contributed by hops is a little higher as the average British hop contains around 0.25 to 1.0% nitrate and American hops can contain levels nearer 2.0%. In Britain the hops usually contribute around 10 ppm and sometimes as much as 20 ppm. Assuming that the average nitrate concentration in the water used for brewing is between 10 and 19.99 mg/l, as just quoted and allowing for a little concentration during wort boiling, the total nitrates in beer should on average be about 30 ppm. This ties in with some British research, in which twenty-five retail beer samples were analysed in 1984-5 and the mean nitrate content was found to be 30 mg/kg, from a range between 10 and 100 mg/kg.

It is of interest to note that hop extracts contain less nitrates than hops, so can be used to significantly reduce nitrate levels in beer. Nitrates can also be removed from brewing liquor by treatments such as ion exchange or reverse osmosis.

Alcoholic beverages are not taken into account in the preparation of Total Dietary Intake figures of nitrate which are published, but even for a moderate drinker of beer the figures are significantly increased when beer is included. The average beer drinker is said to consume 0.7 litres-/day which would contribute another 21 mg nitrate per day. A person who drinks four pints per day (2.28 litres/day) is considered a moderately heavy drinker and the estimated increase would be 68 mg/day. This would more than double the 1979 average daily total dietary intake of 61 mg. It exceeds the total amounts which are consumed by other groups who are regarded as being special. Vegetarians are estimated to be unlikely to consume more than 100 mg/day and extreme consumers of food are unlikely to consume more than twice the average amount of food giving an average daily intake of 122 mg. This will not be of concern to most people but perhaps will be a little more so to the vegetarian who drinks a lot of beer and lives in an area which has a high nitrate concentration in its water.

NITRITES

The concern for the beer drinker over the presence of nitrites is the same as for nitrates which are converted in the gut to nitrites and then react to produce N-nitroso compounds. Fortunately there are not the same level of nitrites as nitrates in food and drink. Water supplies usually contain undetectable amounts of nitrite, but when they do contain any it is usually an indication of sewage pollution, which is of greater concern than the presence of the nitrites. The greatest source of nitrite in a person's diet is from nitrites used as a preservatives in meat products such as pork pies, sausages and bacon. Beer is not a major contributor of nitrite in the diet.

N-NITROSO COMPOUNDS

As well as being produced from nitrates and nitrites in the human body these may also be produced in food during processing and taken into the body directly. This fact has been known since 1972 and since then the MAFF and DHSS have funded research into the nature, incidence, sources and health implications of them in food. These compounds may be formed in food from a nitrosating agent derived from nitrite salts or nitrogen oxides (NOX) reacting with amino substances. Some N-nitroso compounds are carcinogenic and others are not, but one of those found in beer and known to be carcinogenic in thirty-three species of mammals including primates, is N-nitrosodimethylamine (NDMA). When this was discovered in beer through the work of Professor Preussmann in W. Germany, in 1978, the brewing industry was very concerned. It was not desirable that beer should be seen to be likely to increase the risk of cancer, even though there is no direct evidence that NDMA affects man. In order to investigate the situation in the U.K., the Brewers' Society and the Maltsters' Association of Great Britain set up a Nitrosamine Working Party; the MAFF was represented. The monitoring of NDMA was started at the Brewing Research Foundation and in 1980 the number of samples analysed reached 3,000 for the year; the number has dropped to about 1,200 now that the situation is under control. It was found that levels were higher than was desirable, so immediately research into the cause of the high levels and the way in which they could be reduced was started. A confidential report on the situation was prepared and circulated to brewers and maltsters by the working party in 1980, but relatively little was published on the work carried out until 1986. By then there was certainty that the situation was under control and that no health hazard existed.

The source of NDMA in beer was soon identified as being malt, as similar amounts were also found in malt whisky and malt vinegar. A survey of NDMA levels in beer carried out in the U.K. between 1978 and 1979 showed that levels covered a wide range of values up to 20 μg/kg; only 25% of samples were below the level of 0.6 μg/kg. Another survey taken in 1981 showed that levels were generally much lower and not one was more than 5 μg/kg with 70% of the samples falling below 0.6 μg/kg with a median of 0.4 μg/kg.

In the U.S.A. there was similar concern at the time, but the matter was dealt with more publicly and with much more drama. The U.S. Brewers' Association, in 1978, publicly acknowledged the presence of nitrosamines in beer but played the matter down. However, a Chicago television station took up the case and had samples of beer analysed, showing them to be high in nitrosamines. The Food and Drug Administration took their own samples, but were reluctant to reveal their results when requested under the Freedom of Information Act. Eventually after petitioning from the Center for Science in the Public Interest they were published and showed some beers to be over the 5 ppb limit set by the Government. American brewers are perhaps a little more forward about seizing marketing advantages than British brewers and one company which had the good fortune to have low levels in their beers, started an advertising campaign which pointed out in very blunt terms, the high levels in competitors' beers. This resulted in panic in some areas and caused the brewers as a whole, under government pressure, to act very quickly to remedy the situation. They were given three months by the FDA to reduce their levels. This was complied with and levels in beers tested in June 1980 were all below the set limit.

When the diet is considered as a whole, the three foods which are major sources of NDMA are fish, bacon and beer. Other foods such as some cheeses and salami are consumed in such small quantities, on average, that they are not considered large contributors even though they contain high levels. The estimated intake of NDMA from fish, beer and bacon is between 0 and 0.7 μg/day. In the case of the maximum intake: 0.2 μg would be from fish, 0.1 μg from bacon and 0.4 μg from beer, based on average figures of consumption. It can be seen from these figures that heavy beer drinkers would easily increase the percentage of NDMA in their diet. Extreme consumption of beer is more common than extreme consumption of fish or bacon, so on the whole beer must represent the most likely source of excessive levels of NDMA in a diet.

Concern over the levels of NDMA in beer is shared with other countries who have also worked to reduce the levels in their beers. In the U.S.A. for example, twenty-five beers sampled in 1979 had an NDMA mean level of 5.9 μg/kg and a similar sampling in 1981 showed a level of 0.2 μg/kg. Similar results were obtained in other countries.

The occurrence of NDMA is attributed to the reaction of N_2O_3 and N_2O_4 in the flue gases of the malt kiln with hordenine and gramine in the malt. A variety of measures have been taken to lower levels of NDMA. Indirect heating of the malt, cooler gas flames on burners, sulphuring of kiln gases or the use of sulphur-rich fuels have all been used to reduce the effect of the nitrogen oxides. Various measures to reduce the formation of hordenine in malt during malting and the prevention of the nitrosation of hordenine by such measures as lowering the pH of the surface of the malt have also been used.

The presence and levels in beer of the volatile N-nitroso compound NDMA have now been established but less is known about the other N-nitroso compounds, particularly the non-volatile ones which form a large percentage of the total. Recent work has been carried out to determine the levels of apparent total N-nitroso compounds (ATNC) in malt and to find out whether they are affected by the measures used to lower the levels of NDMA. Malts kilned under the conditions where nitrogen oxides (NOX) were low, produced low or undetectable levels of ATNC but sulphuring had little effect in reducing levels.

The ATNC contributed by malt should be less than 20 μg/kg. but recent studies have shown levels as high as 500 μg/kg in beer. These high levels of ATNC have been found to result from microbial contamination of the beer or wort. Some micro-organisms possess nitrate reductase activity and the nitrites produced from nitrates in the wort or beer undergo nitrosation to produce ATNC. The reaction is very quick and not inhibited by heat including wort boiling. This is a further case for reducing the amount of nitrates in beer as well as maintaining the highest levels of hygiene in the brewery.

In a recent study of four brewing yeasts only one did not reduce nitrate. Those that did were contaminated with *Obesumbacterium proteus*. Nitrosation was fairly complete and 2,000 ppb ATNC was produced from the amines.

In February 1987, the Brewers' Society after consultation with the MAFF recommended that brewers should make every effort to ensure that ATNC levels do not exceed 20 ppb. The extent to which this is achieved is not public knowledge but will be more likely to be achieved by brewers with the facilities and knowledge to deal with the situation.

WATER

Water is a major ingredient of beer and offers the greatest potential for the introduction of contaminants. The devastating consequences of even small supplies of water being unsafe through contamination was recognised in the nineteenth century and led to the early introduction of legislation to protect the consumer.

The first piece of legislation was the Waterworks Clauses Act of 1847 which required all water supplies to be 'pure and wholesome'. This was a laudable requirement but somewhat impractical for many purposes, so under the Public Health Act 1936 it became only necessary to supply water for domestic consumption which was 'wholesome'. The word has never been fully defined in this context, but generally means that the appearance, taste and amounts of chemical and microbiological contaminants in water must make it safe and acceptable to drink. Water suppliers are quick to remind brewers that the water which they supply, although it is reputedly potable, is not necessarily suitable for brewing. The meaning of 'wholesomeness', although undefined in British law, is defined to some extent under EEC law. An EEC Directive of 1980 sets limits on about fifty different parameters for drinking water. At the time of writing in 1989 this Directive still has not been adopted by Britain but as far as the brewer and the beer consumer are concerned, it is believed by many that it will offer no benefits.

Wholesome does not mean pure but as long as contaminants are limited the consumer can be sufficiently protected and the cost of the supply can be kept acceptable. However, the sources of contamination are growing and although the technology exists to deal with them, the cost of removing or limiting them will push up the price of water. As the levels and numbers of contaminants increase the risk of them not being contained for one reason or another or the risk of accidental over-additions of treatment chemicals also increases.

It is useful to know the sources and nature of water contaminants which may eventually end up in brewing liquor. The source of all water is rain and with the publicity given in the last few years to acid rain and nuclear fallout it is of no surprise to most people to be told that rainwater is far from pure. Contaminants include dust, soot, pollen, products of combustion, sea spray, and nuclear fallout. When water lands on the surface of this planet more contaminants are picked up. If it falls on fertile soil, nitrates, residual pesticides and organic material are picked up. Nitrate levels are increasing every year as more fertiliser is used in modern intensive farming methods which started after World War II. The problem caused by this was looked at earlier in this chapter. On peaty moorland, organic matter is picked up with some discolouration. On soluble rocks, salts causing water hardness and fluorides are dissolved. These are discussed in Chapter 5 under liquor treatment. On roads rainwater washes off pollutants such as rubber, lubricants, corrosion products, lead from exhaust gases and salt. These can all pass into ground water or surface water supplies.

River supplies also receive water from other sources which may contain treated or untreated effluent. Domestic effluent increases the levels of phosphate and boron. The phosphates come from the food we eat and from detergents and the boron comes from the detergents.

Effluent from industry and agriculture are a source of many other pollutants. There is particular concern over the synthetic and natural hydrocarbons which are in industrial effluent. Some of the more common ones are trichloroethylene, tetrachloroethylene, methylchloroform, methylene chloride, benzine and toluene which are all solvents. Heating and fuel oils are also common pollutants. Apart from the usual levels of industrial pollution, there are also accidents which result in large discharges into rivers. An example of such an occurrence was in

January 1984 when a phenolic substance was discharged into the river Dee affecting 2 million consumers including some breweries. Water suppliers have to monitor river pollution in order to protect their sources. When pollution is discovered the use of the source has to stop until the pollution clears. These types of incidents occur about a hundred times every year when pollution is discovered.

WATER TREATMENT

In order to make water wholesome some chemical treatment in the water treatment works is necessary. Aluminium sulphate is added to precipitate suspended solids and remove the brown colour of some water sources. If water has a low pH it is usually raised by the addition of alkaline calcium hydroxide. This prevents the corrosion of iron pipes and the pick up of iron salts. Polyphosphates and silicates are also used to prevent corrosion. Copper sulphate is sometimes added to reservoirs in summer to kill off algae which cause algal blooms. Lime is also sometimes used to reduce the hardness of water, as in the treatment of brewing liquors which has already been described. Chlorine is used as a disinfectant leaving a residual amount in the water as protection. Sometimes ammonia is added first in order to produce the more persistent monochloroamine. Fluoridation with fluosilicic acid or sodium fluoride is carried out in some areas in order to help prevent tooth decay.

These chemicals used to produce wholesome water can, in some circumstances, become contaminants themselves or produce contaminants. Chlorine in the presence of some types of organic matter such as that extracted from peat and algae produce trihalomethanes, probably as a result of reactions with humic acids. These compounds are believed to be responsible for bladder cancers above certain levels. Trihalomethanes include: chloroform, monobromo-dichloromethane, dibromo-monochloromethane and tribromomethane. The 1980 EEC Directive sets a maximum limit of 100 ppb.

Accidents and human error at treatment works sometimes result in high levels of chemical addition. Most people will at sometime have smelt excess chlorine in their drinking water though sometimes excess has to be deliberately added to eradicate some contamination. Then excess chlorine is hopefully the lesser of two evils. Other excesses can be put down to miscalculation or accident. The author has experienced high levels of calcium hydroxide in a mains water supply which raised the pH above 9, making it an unacceptable effluent according to the same water authority's requirements. More serious mishaps have occurred such as when 20 tons of aluminium sulphate were emptied into the water supply at the Lowermoor treatment works, near Camelford, in North Cornwall, in July 1988. Health problems in the people who had drunk the water became evident in 1989. The symptoms experienced include arthritis, memory loss and skin rashes. The experience has also highlighted recent information linking high levels of aluminium in the diet to Alzheimer's Disease (senile dementia) and the fact that up to a million people in Britain have levels of aluminium in their drinking water which is above the maximum level in the EEC Directive.

Serious incidents are fortunately rare and water suppliers are generally keen to emphasise the high quality of water in Britain. However, the brewer has to regard it as a potential source of contamination of his product. Apart from the adjustment of the mineral salts as described under the section on liquor treatment, some brewers take steps to totally treat the water. This first involves ultra filtration if this has not been carried out adequately by the supplier, in order to remove suspended solids and colloidal iron deposits. This is followed by complete demineralisation which will remove all minerals including the undesirable nitrates. Carbon filtration will then remove most of the other contaminants not removed by the demineralisation. The final step is sterilisation with ultra violet light.

The likelihood of contamination will depend on the approach of the brewery to the current situation. It will vary from the total treatment approach with monitoring of the purity before and after treatment, to the situation where a private supply is used which does not even have the benefit of the monitoring carried out by water authorities.

PLASTICS

Plastics are inert, stable polymers but during the manufacture of plastic items they are mixed with additives such as catalysts, plasticisers, processing aids, stabilisers, and pigments. They may also contain the residues of the monomer from which the polymer is made. These substances can cause problems when a manufactured plastic article comes into contact with food if they migrate from the article to the food.

An example of a substance which caused a problem in beer is sulphonamide which is used as a plasticiser in nylon beer piping. It leached out of the piping into the beer, so that with every pint of beer dispensed, a small dose of sulphonamide was included. Such cases lead to the introduction of the Materials and Articles in Contact with Food Regulations being introduced in 1978.

The part relevant to plastics in contact with beer states:

> Materials and articles to which this regulation applies shall be manufactured in accordance with good manufacturing practice, that is to say in such a way that under normal foreseeable conditions of use they do not transfer their constituents to foods with which they are, or are likely to be, in contact, in quantities which could:
> 1. endanger human health, or
> 2. bring about a deterioration in the organoleptic characteristics of such food or an unacceptable change in its nature, substance or quality.

The regulations also prohibit any person from selling or importing materials or articles which do not comply with the above part of the regulations.

BEER LINES – SMALL BORE

Plastics frequently come into contact with beer before it reaches the customer's lips and perhaps the greatest contact time that a beer will have with plastic is in the dispense line. The two most common plastics used for small bore dispense piping are polyethylene (polythene) and nylon. In the case of low density polythene it can be used without the addition of any other chemical substance as its natural characteristics make this unnecessary. The only other substance likely to be present will be unconverted monomer which has not been polymerised during manufacture.

The other type of plastic used is nylon 11 which is relatively inflexible. It is used with the plasticiser which has already been mentioned, but since the legislation introduced in 1978, nylon piping has been produced with a lining of unplasticised material in order to prevent plasticiser passing into the beer. This is done by coextrusion. Even so there have been cases reported of beer being tainted when dispensed through this tubing.

BEER LINES – LARGE BORE

If the beer is dispensed through beer engines large bore piping has to be used and this is usually made with PVC. Problems of contamination with the toxic monomer vinyl chloride from which the polymer is made led to the introduction of EEC legislation limiting its presence in the polymer to 1 ppm. The problem originally became evident in the U.S.A., about 1973, when vinyl chloride was discovered in distilled spirits packaged in PVC bottles. The FDA immediately proposed to withdraw approval of the use of PVC as a suitable material for packaging materials in contact with alcoholic foods. In Britain the MAFF, rather than making any hasty decisions concerned itself with gathering more information. In 1974 the FACC decided that they should encourage the levels in food to be reduced, but that it was premature to recommend any further action. The amounts consumed have been steadily reduced since then. The maximum per capita intake in total diet studies in the U.K. was 1.3 μg/day in 1974 and by 1978 had reduced to 0.02 μg/day. However, these figures have little value for beer drinkers as beer was not included in the total diet studies.

On its own, PVC is hard, unworkable and prone to degradation. The main additive used is a primary plasticiser which gives the material flexibility. There has in recent years been concern over the use of plasticisers which lead the MAFF to gather information on their use, the levels found in food and dietary intake levels. Over a hundred plasticisers are in current industrial usage. In 1984 the total annual usage of all food contact plastics in the U.K. was 5,000 tonnes of which 500 tonnes was PVC used in beer lines. Plasticisers in PVC are used at the level of 30%. The main ones are di-2-ethylhexyl phthalate (DEHP) and di-iso-octylphthalate (DIOP).

The Steering Group on Food Surveillance (sub group on plasticisers), examined the use of plasticisers and published their report in 1988 in the Food Surveillance Paper No. 21. They had nine samples of draught beer analysed. The levels of DEHP and DIOP, which were measured together, were in the range <0.01 to 0.04 mg/kg with an average of 0.03 mg/kg. Di-2-ethylhexyl adipate (DEHA) was also found to be present at levels between <0.01 to 0.07 mg/kg with an average level of 0.02 mg/kg. This plasticiser is normally used in PVC film wrappings and its presence in the beer analysed could not be explained by the Group. The Group also understood that beer which had stood in beer lines for any length of time was disposed of, as it lost carbonation. This is an assumption which cannot always be safely made. Plasticisers were also found in bottled beer where they had migrated from plastic used in closures. Results showed levels below 0.02 mg/kg. The Group reported, using diet studies, that the maximum daily intake of DEHP and DIOP is 0.02 mg/day calculated on the basis of their presence in beverages. However, this seems a little out of line with the levels found in beer and some daily consumption figures. In relative terms this is still not high compared with plasticiser levels in some other foods brought about by the use of film wrappings

The COT and FAC also published their views on the Group's findings in the Food Surveillance Paper. The COT reported that levels of DEHP were 100,000 times lower than the level at which the first indications of toxic effects on rodents were noted. Therefore, continued use at present levels was considered to carry no risk, though it was thought that usage levels should not be allowed to increase. However, they stated that there was a 3,000-fold margin between levels of usage of DEHA and levels which have produced carcinogenic effects in mice and that in this case levels of usage should be reduced.

Degradation of PVC is prevented by stabilisers which are often metallic soaps such as zinc laureate and zinc stearate. An antioxidant such as nonyl penyl phosphite is also usually used as well as a secondary plasticiser such as epoxidised linseed oil. A lubricant such as stearic acid is used as a processing aid to help the extrusion of the plastic in manufacture. Some of these substances leach out of the plastic into the beer but it has also been shown that part of the surface of PVC tubing is physically removed when fluids pass along it, so small amounts of these substances are almost certainly consumed.

Another problem with plastic additives is that they can react with some sterilants and cleaners to produce compounds which, if not thoroughly washed off, can produce taints in beer. An example is chlorine which can react with phenolic compounds to produce chloro phenols which have an exceedingly low taste threshold value.

Another plastic which has been used in the manufacture of beer dispense lines is ethyl vinyl acetate (EVA). It is flexible and could be used as an alternative to PVC but unfortunately it can cause taints so is not frequently used.

BOTTLES AND CONTAINERS

Another occasion when plastics come into contact with beer is in the use of plastic bottles. PET (polyethylene terephthalate) bottles for containing beer were introduced in 1981. The plastic is considered safe and in the U.S.A. has FDA approval. No additives or plasticisers are used but the bottle has to be coated on the outside with a thin layer of vinylidine chloride copolymer (PVdC) in order to provide a non-porous barrier which stops air getting into the bottle and carbon dioxide getting out. This prevents the bottles being recycled. Pigments have now been added which prevents the production of light-struck flavour and specifications for this have been approved with the Brewers' Society.

Other plastics such as PVC have been used for the manufacture of containers used for the temporary transport of beer from the retailer to the home, but of course they suffer the same problems of toxicity encountered with PVC pipes.

Before the introduction of PET, acrylonitrile was used for making disposable plastic bottles. It was used for a number of years but the high cost of acrylonitrile stopped production. It has also been known for a number of years that acrylonitrile is toxic and it is now accepted that in sufficiently high doses it is carcinogenic to man. A working group of the Steering Group on Food Surveillance collected information on the use of acrilonitrile, the levels found in food and dietary intake levels. Details of their findings were published in the Food Surveillance Paper No. 6, in 1982. Beer was not seen as a food likely to contain high levels of acrylonitrile and was not even mentioned, although acrylonitrile bottles were in use when investigations were made. Soft margarine packed in acrylonitrile containers was of the greatest concern but even levels in this were well below those at which there is any risk to health.

MISCELLANEOUS USES

Plastics are also used in other parts of beer dispense systems. Acrylonitrile, butadiene styrene co-polymer and acetal polymers are being used to make items such as cask taps, valves and parts of beer engines.

CONTROLLING THE PROBLEM

The Brewers' Society has made its members aware of the problems of migration of substances from plastic articles. They have advised on the methods of checking plastic equipment for its effects on the quality of beer but these tests are to check for effects on flavour and not to detect levels of toxicity which may affect customers. In the Materials and Articles in Contact with Food Regulations 1978 provision was made for the testing, examination and analysis by public analysts, but such tests are basic and do not necessarily include all the conditions under which the materials may be used in beer dispense. However, there are analysts available who can conduct tests under conditions specified by the customer. Advice can also be obtained from the MAFF or BIBRA (British Industrial Biological Research Association) through the Brewers' Society. In addition a Code of Practice for the safe use of plastics has been jointly published by the British Plastics Federation and BIBRA.

PESTICIDES

In recent years there have been developments which have improved the level of protection of the consumer from pesticides. The Control of Pesticides Regulations 1986 and an enquiry into pesticides and human health conducted by a House of Commons Select Committee in 1987 are the two main developments. However, many consumer and environmental organisations are not happy with the situation in the U.K. and feel that the country lags behind other developed countries.

As far as the beer drinker is concerned, the two main ingredients of beer are both treated with pesticides. The use of pesticides on barley stored for malting have been mentioned in Chapter 5 but pesticides are not used on hops stored for brewing. However, both ingredients are treated with pesticides by the farmer during cultivation. The main concern in recent years has been over the intensive use of pesticides on hops. Hops are very susceptible to attack from a variety of pests and because they are such an expensive crop to cultivate the hop grower has to draw on all the resources available to protect his crop.

The fight against the excessive use of pesticides on hops has been led by the U.S.A. and under their law 95% of the hops grown in the U.K. would not be acceptable for use in the U.S.A. In Czechoslovakia, where they are aware of the problem with pesticide residues, 2,000 zentners were rejected in 1987 because of pyrethroid contamination. In 1988 30,000 zentners (10% of their production) were rejected because of bromopropylate contamination. In Germany they have been analysing their hops for thirty three different residues since 1986 but in the U.K. the analysis of hops for residues is far from common practice at the time of publication.

FILTER MEDIA

When beer is processed in the brewery any yeast and protein-tannin material in suspension is removed to produce a bright beer. The bulk of material is allowed to settle on the bottom of the conditioning tank which may be assisted by the use of finings or other processing aids. Sometimes material is removed from the beer by means of a centrifuge but ultimately any remaining material is removed by filtration. There are basically two types of filters: powder filters and sheet filters.

In powder filters a filter bed of a suitable powder is built up on a supporting medium and the beer is pushed through this under pressure. The solid material which is removed would soon blind the filter, so powder is injected into the beer as it goes into the filter. This keeps the solid material porous when it is deposited on the filtration surface.

Filter sheets are thick sheets of compressed fibres which are self supporting. They are used to remove smaller amounts of solids which are left after powder filtration or centrifugation. They can be re-used if the solid material deposited is removed by backwashing.

KIESELGUHR

This is the siliceous skeletal remains of diatoms or diatomites which lived in seas and lakes twenty-one million years ago, in the Miocene Period. Deposits of their remains are now above sea level in Europe and America where they are mined. After the material has been mined it is ground, sterilised, calcined at between 800 to 1200°C and then classified into different grades. The purpose of the calcining is to remove organic contamination.

Chemically, kieselguhr is hydrated silica, which is similar to opal. It is very inert which makes it suitable for filtration, but it is very susceptible to contamination by taints which can be passed on to the beer. Some years ago a consignment of kieselguhr was shipped with some oranges. Their flavour was picked up by the powder and passed on to the beer which it was used to filter.

Kieselguhr was first used in the U.K. around 1930 and has remained in popular use in powder filtration since that time. At one time its action was enhanced by the addition of asbestos fibres but these are not now used.

PERLITE

This is a volcanic deposit obtained from certain Greek islands. It is cheaper and safer for operatives to use than kieselguhr but is not widely used.

PULP

Pulp, produced from pure cotton fibres, was introduced as a filter medium at the end of the nineteenth century. The fibres were formed into pads which were fitted into a filter press. They were broken up after use, washed in mild caustic soda and bleach and then re-used. Like other filter media asbestos fibres were also added in order to improve performance.

SILICA HYDROGELS

These have been used for a number of years to stabilise beer in conditioning tanks but in the 1980s they have been used to replace kieselguhr in powder filters.

CELLULOSE ACETATE AND NYLON 66

These are used in membrane filters which are sheets of inert material made with a specific pore size. They act like minute sieves and easily blind, so they are only used as guard filters which act as a back up in case there is a failure of the main filter.

ASBESTOS

Sheet filtration was introduced around 1930 and originally the sheets were made of mixtures of asbestos, kieselguhr and cellulose. Until the 1980s asbestos was the main material used but its popularity has declined because of safety fears.

There is a well-known relationship between the breathing in of asbestos dust and asbstosis, asbestos-related cancer and mesothelioma. The publicity given to the effects of breathing in asbestos dust made some people fearful that ingestion of minute traces of asbestos in food and drink would cause slow and long-term damage to the intestine. Several reports on research carried out in the 1960s and 1970s showed that there is no risk from ingesting asbestos fibres, but a report from the U.S.A. states that the rate of cancer of the stomach, intestine and rectum is three times higher in asbestos workers than the rest of the population.

Some attention turned to beer, because of the use of asbestos fibres in filter sheets. The normal risk of fibres passing into the beer was low unless normal filtration procedures were abused. Tests carried out showed that, in some beers, the fibres present before filtration were greater in number than after filtration. This can be explained from the fact that there are thousands of miles of asbestos-cement pipes used in the water supply industry, which contribute asbestos fibres to the water which runs through them.

Though the risks to beer drinkers are believed to be negligible, rather than run the risk of adverse publicity most brewers have chosen to use the non-asbestos filter sheets which are now available. The advice of the FACC in their report on additives in beer, in 1978, was that although they considered that there was no danger from asbestos an alternative should be found. This has perhaps been good news for the brewery workers more than anyone else as it was they who had to handle the filter sheets which put them at most risk.

Water supplies and filter sheets are not the only sources of asbestos fibres in beer. Research in the U.S.A., in 1975, revealed that in most cases the numbers of asbestos fibres in bottled and canned beer was higher after the filling operation. In the case of cans the extra asbestos appeared to arise from the asbestos gaskets of the drying ovens used in the process of lining the cans. Fibres also entered during transportation.

All filter media used are inert and present no danger of contamination to the beer filtered through them. Asbestos, because of the nature of its fibres, is the only one which is likely to give the consumer cause for concern. The Food Additive and Contaminants Committee in their report in 1978 approved the use of all the other filtration media providing the silicates passed on to the beer did not exceed 2,000 ppm.

TAINTS AND OFF-FLAVOURS

Sometimes beer can become contaminated with very small amounts of compounds which do not necessarily endanger the health of the consumer, but can affect the flavour of the beer to some degree. These can be present through bad brewing practice, accidents, neglect, use of poor quality materials and a variety of other reasons. Many taints are caused by microbiological spoilage but as already mentioned, the subject is outside the scope of this book. Some examples of taints due to other influences are given.

CHLOROPHENOL AND PHENOL TAINTS

These taints in beer are commonly described as medicinal or antiseptic taints. The chemicals 2,6-dichlorophenol and 2- chlorophenol can be detected at very low levels in beer, even as low as 0.5 ppb. Their presence in beer most commonly originates from the action of chlorine-based sanitisers reacting with plastic dispense equipment during cleaning. They are difficult to rinse off and are picked up by the beer when the equipment is put back into use. Research has shown that the plastic, Nylon 11, used in some dispense lines can react with chlorine in such a way. It has also been discovered that dispense tubing affected can be restored by washing it with a solution of sulphur dioxide in isopropanol.

The taint can also result from using water for brewing which has become tainted by chemicals, or by micro-organisms which produce phenolic metabolites.

Some highly cured malts and some varieties of hops have naturally high phenol levels and if a high proportion of these are used, a taint can occur in the beer produced from them.

Other sources of the taint have been attributed to accidents with cleaning materials, raw materials contaminated with pesticides or insecticides and uncured lacquer on cans and kegs.

In some cases the taint has originated in steam which has leaked into wort. Chlorophenols are sometimes present in perceptible amounts because they have been concentrated in the boiler feed water or have been generated from chlorine residues.

CHEMICALS PRODUCED BY ELECTROMAGNETIC RADIATION

It is well known that beer in clear glass bottles is affected by ultraviolet light, a wavelength which occurs at particularly high levels in bright sunlight. It produces a sulphury flavour which is known as light-struck or sun-struck flavour. The chemical which gives the flavour is 3-methyl-2-butene-1-thiol and is produced by the action of the light on the bitter isohumulone side chains. Most beers for this reason are bottled in dark glass, but some dark strong beers are less susceptible and can be safely bottled in clear glass. The effect of artificial light varies with the source and intensity but in all cases high pressure sodium light has the worst affect.

It is worth mentioning the effect of irradiation by gamma rays as this can be used in some countries for sterilising food. It is a satisfactory method for some wines, but is totally unsuitable for beer because of the production of compounds with off flavours. Their production is thought to involve proteins which are at a much higher level in beer than they are in wine. Canadian workers reported that it gave their beer a flavour which was a mixture of sour, burnt, pungent and sun-struck flavours. Belgium researchers reported a flavour in pilsener beer which was a mixture of hydrogen sulphide aroma, sun-struck flavour and autolysed yeast flavour.

METALLIC FLAVOURS

These flavours are also described as inky or drying flavours and are mostly due to the presence of iron. This can get into beer from a number of sources, particularly where old and badly maintained plant is used. Cast iron liquor tanks were always whitewashed to prevent the pick up of iron, but if it was not regularly carried out the protective surface broke down. Cast iron yeast presses were also regularly painted with tannic acid to give them a black protective coating. Tanks and beer filters of mild steel were always manufactured with an enamel, resin or glass coating but this could easily be chipped and if not repaired, exposed the beer to iron. The increased use of stainless steel has somewhat overcome this problem but if inferior grades are used, the problem can still exist. New stainless steel always needs to be pickled in detergent and some waste beer before it is used in order to avoid picking up a metallic taste in the beer. This is probably due to minute but loose particles of metal left on the surface after polishing. This advice incidentally applies to all new brewery equipment including rubber and plastic hoses.

Another source of metallic flavour can be from cans where the protective lacquer has broken down. On seamed cans this was usually at the seam, but elsewhere on the can it occurs as small pin holes in the lacquer.

DETERGENT, COOLANT AND LUBRICANT CONTAMINATION

Brewing is no different from any other human activity and human error or mechanical failures have resulted in accidents in which chemicals become mixed with the beer. Plant in a brewery has to be thoroughly cleaned and sterilised each day using detergents, many of which are based on caustic soda. Incomplete rinsing of the detergent or accidental additions from the cleaning systems have occurred in breweries at some time. Caustic soda in beer gives it a distinctive sickly, biscuit-like aroma, makes it feel soapy and affects clarity.

Leeks from attemperators and other cooling systems in contact with the beer can occur and coolant (secondary refrigerant) such as calcium chloride or alcohol can pass into the beer.

Lubricants on bottling and canning lines have been known to contaminate beer. There was even a case of lubricant used in the riveting of the rings on ring pulls getting into beer and contaminating it.

CONTAMINANTS PRODUCED BY MISMANAGEMENT OF FERMENTATION AND CONDITIONING

Dimethyl sulphide is one of the main compounds which give lagers their distinctive flavour. However, excessive amounts can be found in beer through mismanagement of conditioning. Excessive amounts are normally washed out of the beer with the evolution of carbon dioxide, but mismanagement can stop this occurring.

Esters and fusel oils, which have very strong and distinctive flavours, can be produced in excessive amounts if the conditions of fermentation, such as oxygenation and temperature, are not properly controlled. Examples of unsatisfactory flavours produced in such a way are those which are described as fruity and solvent-like.

Sulphury flavours can be produced by allowing excess trub to pass from the copper into the fermenting vessel.

The level of diacetyl, which has a buttery flavour, can be affected by the composition of the fermenting wort and by the speed of fermentation.

Yeast bite, which can be described as an astringent bitterness, can result from a badly managed fermentation and the poor handling of yeast and yeast pressings.

OXIDATION FLAVOURS

The main flavour produced by oxidation is the stale flavour often described as a cardboard flavour. Various compounds in beer are believed to be oxidised and produce unsaturated aldehydes, particularly trans-2-nonenal. In bottled beer oxidation can produce a tom cat flavour, which disappears as oxidation continues, to produce a compound with a sickly, toffee-like flavour. Pasteurisation flavour is another oxidation flavour produced when beer is pasteurised in the presence of air and has been described as a bready, cooked or grainy flavour.

UNUSUAL SOURCES

Some taints in beer have come by very obscure routes and have taken some detective work to find their source. A ribes or cat flavour, once detected in a beer, was traced to the malt. Paint used in the maltings was contaminated with an impurity called mesityl oxide. This reacted with naturally occurring hydrogen sulphide in the malt to produce a compound with a very strong ribes flavour, even at very low concentrations. Another beer which tasted of oranges was traced to the filter media which had been shipped with a load of oranges. The filter media had picked up the aroma and passed it to the beer on filtration. A disagreeable odour in a bottled beer shipped to the tropics was traced to naphthalene formed from the plasticiser in the plastic liner on the crown cork when it had been affected by the high temperatures.

Fortunately these contaminants are only encountered by consumers on very rare occasions. Such flavours are very pronounced and are spotted before the beer ever leaves the brewery, especially today, when quality control in breweries has to be thorough.

MYCOTOXINS

Poisonous substances may be found in food when they become infected with bacteria or moulds which produce toxins. The toxins produced by moulds are known as mycotoxins and have been found in cereals and commodities such as groundnuts and cotton seed, which are used in animal feeds. Toxins in animal feeds have been passed on to man through cows' milk. One of the oldest mycotoxins known is that which causes St Anthony's Fire or Ergotism from an infestation in rye. There is a record of an outbreak in the Middle Ages and various other occurrences have been reported over the ages, until the last major outbreak in England in 1928. Haemagglutenin toxin is another which occurs in red kidney beans which are sold as a natural wholefood. It is destroyed by cooking for 10 minutes but slow cooking at a low temperature increases the amount of toxin.

Fortunately for beer drinkers mycotoxins do not present a problem. A recent investigation in Italy of twenty-four beers, including beers imported from eleven different countries, showed that of the eleven mycotoxins which were looked for, none was present in any amount which could be detected. However, it is of interest to mention that *Candida albicans*, a wild yeast which can live in beer, can also live in the human gut encouraged by diets rich in sugar where it releases a mycotoxin. The effects of this mycotoxin on the body have been studied by Dr William Crook in the U.S.A. He has found that the mycotoxin weakens the immune system producing symptoms such as fatigue, headaches, skin disorders, numbness, sore throats, aching muscles and hyperactivity. He has successfully treated the condition.

CHAPTER 7

ADJUNCTS

ADJUNCTS BEFORE 1880

In Britain, malted barley has always been the principal source of extract used in brewing, but from early times other materials have been used for various reasons. The Celts used honey in some of their beers, though whether its use was primarily as a sweetener or as a source of extract one can only guess. Mead was one of the other beverages known to have been produced at the same time, so perhaps honey flavoured beer was a hybrid beverage which suited those who were accustomed to the flavour of mead. In Mediaeval times other cereals were used for brewing in various parts of the country. Availability and economic considerations must have first prompted their use but once a following for the beers in which they were used had been established, this must have been the major influence on their continued use.

WHEAT

It is not technically possible to produce an all-wheat beer but up to half of a malted barley grist can be replaced by wheat. Beers brewed in this way were known to have been produced in Mediaeval times. In some parts of the country oats were also used in the grist with the wheat. However, wheat was generally reserved for baking and was even banned for brewing on occasions so that the price of bread was not forced up. Later in the sixteenth and seventeenth centuries special wheat beers were being produced, particularly for export. When malt was taxed in 1697, the use of wheat for brewing was prohibited and it was not used again in any significant amounts until relatively recent times.

Perhaps rather surprisingly one of the countries where wheat has always been widely used is Germany, where they choose to apply so-called beer purity laws. Their wheat beers seem to be convenient exceptions. One city noted for its wheat beer was Nuremberg. It had regulated its brewing industry since the fourteenth century, permitting only the use of malted barley, but at the beginning of the seventeenth century wheat was allowed to be used. This was because in 1541 Dutch wheat beer was introduced and became so popular that it began to affect the sales of the city's red beer. In Hamburg records of 1458 show that 10% of the beer was produced from wheat, but by 1575 most of it was wheat beer. The beer was described as the Queen of Wheat Beers and production had been turned over to it because of the popularity of imported wheat beers from England. In Lubeck in the sixteenth century there were two types of brewers: the white wheat beer brewers and the red barley beer brewers.

Several famous German beers were wheat beers. Einbeck beer was perhaps the most famous beer in the world in the Mediaeval period, being widely exported, even reputedly as far as Jerusalem. It was brewed with a third part wheat. The well-known Brunswick Mumme was a wheat beer. Broyhan, a beer from Hanover, first brewed in 1526 in an attempt to copy Hamburg wheat beer, brought the town prosperity and was copied in many other towns. Wheat beers became so popular that even Munich, famous for its dark beers brewed with malted barley, started brewing wheat beer in the sixteenth century.

OATS

Oats have been used since early times for brewing in Britain. In Mediaeval times malted oats were used in the north and west and elsewhere mixtures of malted barley, oats and wheat were common. A mixture of malted oats and barley was known as dredge malt in the late Mediaeval period. Oats were not used where there was a choice. Andrew Borde, in 1582, wrote: 'Barley maketh better ale than oaten malt or other corn dough.' However, there was often no choice.

In the eighteenth century it was recorded that malted oats were used in the wet upland areas of Cheshire, Derbyshire, Lancashire, Devon and Cornwall where barley was scarce. Scotland is noted for its production of oats so it is not surprising that in the eighteenth and nineteenth centuries, Edinburgh produced a widely acclaimed oat ale. In the same period Yorkshire was also noted for its oat ale.

ROOT CROPS

Parsnips were used for making beer in Ireland in the nineteenth century but have not been used or even considered for use in more recent times.

The brewing of potato beer has been recorded in the past in Norway and America.

PEAS AND BEANS

Morrice, at the beginning of the nineteenth century, said that he had known peas to be malted with the barley at the rate of a peck to every five quarters and that they gave beer a mellow flavour. He also said that beans were used but did not mix so conveniently as they germinated more slowly. J. Death also records that the Ancient Egyptians used dholl, a small red pea, in brewing their beer which was called hek.

HONEY

The only sweetener that was known for many centuries was honey. In Celtic society the rich were said to drink wine and the poorer classes were said to drink beer. Essentially beer was of two types: that brewed with honey and that brewed without honey. This view is supported by a document issued from the convent of Corvey in 1025. In it, Conrad, King of the Germans, confirmed certain rights and revenues granted to the Nunnery at Herford in Westphalia. Among the revenues were specially mentioned those of mead, beer prepared with honey and beer prepared without honey. The sweet honey beer was known to the Greek writers as Korma.

It is of no surprise that the noted Celtic beer of the British Isles, the Welsh beer Bragawd or Bragot, incorporated honey as an ingredient. Honey was not widely used in English beers, though a nineteenth century recipe for Windsor ale includes it. There was some occasional use by private brewers but very little use by commercial brewers.

MOLASSES AND TREACLE

In the nineteenth century treacle and molasses were more popular in Britain than honey. Treacle was used in recipes for Welsh ale, Burton ale, Scurvy grass ale, table beer and porter. It enriched their flavour and supplemented the extract. There was even a treacle beer made without malt which was flavoured with fruit and spices as desired. Molasses was not so popular but its use was recorded in amber and scurvy grass ale. A molasses beer was produced but the following case of its ill effects in the eighteenth century is quoted:

> Notwithstanding this molasses beer got a great name by pleasing many of the ignorant town people, and even tempted the very drawer to drink frequently of it, yet at last this drawer found the ill effect of it by woful experience; for, by drinking a whole quart of it to his own share in a very little time, it presently threw him into a violent sweat, upon which directly ensued a great cold and hoarseness that held him four months, and had like to have kill'd him as he himself declared.

However, in the nineteenth century, another author considered that it made good beer when used with malt and it was used in years when the barley harvest failed. This was only allowed with the permission of the treasury so that the price of beer could remain stable. Private families could of course use it as they wished. A type of molasses beer was also brewed in America and flavoured with ginger.

THE MALT TAX

Malt was fist taxed in 1660, in the reign of Charles II, but no account of the revenue received has been preserved of an earlier date than 1697, according to the 24th Annual Report of the Commissioners of Inland Revenue. The introduction of a tax on malt in Scotland and Ireland came much later in 1713 and 1785 respectively. It was in the interest of the Government, in order to collect the maximum amount of revenue, that materials other than malt should not be allowed for the purpose of brewing, which was mainly the reason for the legislation passed in 1689, 1713, 1802 and 1816, the details of which have already been given. Although this legislation protected the revenue, it did prevent brewers taking advantage of the sources of extract mentioned above and also any new sources which became available. Ultimately there had to be a relaxation in legislation, which did occur during the nineteenth century.

RAW GRAIN

Brewers in the eighteenth century knew that a certain amount of raw barley grain could be used with malt. It had been deduced from the observation that when poorly modified malt was made, in which a number of the grains had not germinated, that most of the starch was still converted in the mash tun. This was quite easily seen as the endosperms of these corns were not left intact in the spent grains. The use of raw grain was illegal but there seems to have been little attempt by brewers to use it on an illegal basis. The producers of distilled spirits, who were in a position to use it legally, recognised its usefulness. They used different mixtures of grain, though barley always predominated. In Northern Europe the use of potato starch in distilling was well known, as it is from this that vodka is produced.

Nevertheless, in Britain in 1802 and 1803 there had been a serious awakening of interest in the use of raw barley for brewing, because of the sharp rise in the malt tax. This rose from 1/4¼d per bushel in 1801 to 2/5d in 1802 and 1803, then in 1804 it rose again to 4/5¾d at which high rate it remained until 1816 when it dropped back to 2/5d. Raw grain was an illegal ingredient, but like the use of other illegal ingredients its use was tolerated at the time. Shannon, in his *Practical Treatise on Brewing*, in 1805, recommends its use to his readers. However, in 1811 the Excise decided to clamp down on its use. It had only been used by the more scientifically minded brewers who reckoned to save up to two thirds of the malt duty. Despite the efforts of the Excise raw grain continued to be used illegally.

Hayman, in 1823, was recommending its use in his *Art of Brewing* at the rate of 20%. Eventually practical measures were taken to prevent its use in 1856, when brewers were compelled to use smooth rolls on their mills. The hard raw grain skidded on the smooth rolls and could not be crushed. However, with the successful use of sugar in brewing the use of raw grain as an adjunct was pursued very little after 1870.

SUGAR

Failure of the harvests in 1800, 1812 and 1813 lead Parliament to quickly pass legislation which allowed the temporary use of sugar in order that there were adequate supplies of beer. The maximum which could be used was 25% replacement.

The next concession, made by Act of Parliament in 1847, was to allow the use of sugar on a permanent basis. The main reason for this was to put Colonial produce on the same footing as the British product on the home market. When the Act was passed the customs duty on sugar was rather more than the equivalent tax on malt, so no further taxation was levied upon it. When the Customs duty was reduced in 1850 it became necessary to charge a countervailing duty on the sugar used for brewing, making it the same as the equivalent tax on malt. This worked out at 11/6d per cwt against 21/8½d per quarter of malt. The weight of sugar taken as being equivalent to a quarter of malt was initially 200 lb, but was then revised and set at 210 lb. British sugar, which was a very small proportion of the total, was made from either beet or cereal starch and charged at the same duty rate. The largest amount of British sugar on which duty was ever paid was 83,690 cwt in the year 1873–4. Of this 75,000 cwt was made from starch and 8,000 cwt from beet.

The amount of sugar used, until 1866, never exceeded 10 million pounds. Subsequently its use increased greatly and in the year 1879–80 it exceeded 127 million pounds. This was attributed to a reduction in the prime cost of sugar and because sugar refiners were producing a better product. The number of patents taken out by Manbré, Garton and Richardson in the 1870s for improvements in the production of sugar confirms that there was considerable effort being made to improve production methods.

Initially sugar was merely seen as a replacement for malt, but the effect it had on the flavour of beers led brewers to introduce new beers in which the luscious flavour of the sugar was a more pronounced feature. Primings, though they were introduced to give a quick maturation to running beers, also imparted their luscious flavour to the beer. As their use increased, running beers generally became more sweet and it accentuated the difference between the running beers and the stock beers, probably hastening the decline of the later.

OVERSEAS SUGAR INTERESTS

One of the people who strongly petitioned the Government to allow sugar to be used in brewing and distilling was Benjamin Greene, the original Greene of Greene King, the Bury St Edmunds brewers. He was a witness before the House of Commons committee investigating the use of molasses in breweries and distilleries, in 1831. It was only when they were three-quarters of the way through his evidence that he admitted to an interest in the matter, in the form of a large sugar estate in St. Kitts, which he had acquired in 1825. Molasses, however, was of limited practical use, being only of value as a colourant for porter. This fact coupled with the opposition of landowners in Britain, who saw their barley sales under threat, led the committee to recommend against its use. When sugar was eventually allowed for use in 1847, it was too late and with too little potential to help the already failing fortunes of the Greene's sugar estates.

SUGAR FROM GRAIN

In the early years of the nineteenth century scientists tried to chemically reproduce in the laboratory the changes which occurred with natural enzyme systems in the mash tun. Eventually it was a Russian scientist, Kirchoff of St Petersburg, who chemically converted pure fecula (potato starch) into a sugar substance. His work, published in 1811, showed that he used sulphuric acid and a long period of boiling in order to achieve conversion. Scientific work in this field continued on the Continent and perhaps more so in France than anywhere. A short time after Kirchoff's work was published, de Saussure showed that the reaction was a hydrolysis and that the substance produced was identical to grape sugar or glucose. Dubrunfaut then applied this scientific discovery and on a large experimental scale produced sugar from potato starch, from which in turn he produced a distilled spirit and a type of beer.

In Britain, industrial production of sugars from starch began in the 1840s with the manufacture of crude chip glucose. C. Farina had taken out a patent as early as 1837 for producing fermentable matter from grain. Developments continued and in 1859 A. Manbré had patented: 'An improved method of extracting and purifying sugar called glucose from potatoes, or starch, or dextrine, for the purposes and uses of brewers, etc.' M. A. F. Mennons followed with another patent for the manufacture of glucose from vegetable products in 1862. Other patents then continued to be taken out fairly regularly including ones in 1864, 1877, 1878, 1879 and 1880.

THE CUSTOMS AND INLAND REVENUE ACT 1880

The biggest influence on the use of adjuncts in Britain was the Customs and Inland Revenue Act of 1880, which completely changed the taxation system in connection with beer. The malt and sugar duty were abolished and duty on beer, which had been abolished in 1830, was reintroduced. The new beer duty was, however, made payable on a completely different basis. Instead of being levied on the finished product, it was levied on the strength (original gravity) and the volume of the wort collected during brewing, from which beer is produced. This was only then possible because of the greater availability of accurate saccharometers used to measure specific gravity. Because the tax on malt had been abolished there was no advantage in prohibiting substitutes in order to protect the revenue. The only barrier to allowing their use was whether it would have a detrimental effect on the economy and whether it was in the interests of the health of the consumer.

The Government decided for various reasons to allow the use of substitutes and, as Gladstone put it, 'to free the mash tun'. The main reason given for the change was that the malt tax was 'a tax on a raw material and not in accordance with the recognised principles of political economy'. It was also stated that the malt tax imposed 'restrictions on manufacturers which custom only made tolerable and that it necessitated legislative interference in every stage of manufacture'. A writer in 1900, looking back on the changes of 1880 declared them to be the 'greatest fiscal change accomplished in modern times and a measure which many had desired but which few had deemed practicable before 1880'.

ADJUNCTS AFTER 1880

Today there are a wide variety of adjuncts available to the brewer and good use is made of them in Britain. They are broadly divided into the following groups:

1. The starch-rich adjuncts which are added to the mash tun with the malt. They are mostly cereal based though some of the lesser used adjuncts are obtained from plant roots or tubers.
2. The sugar-rich adjuncts which are obtained from sugar cane or beet, and more recently from cereal or root starch. They are also sometimes known as copper adjuncts as they are added to the brewing copper. They do not need to be chemically broken down in the mash tun like the starch-rich adjuncts. Some are also used as priming sugar which is added to the beer after fermentation.

LIMITATIONS OF USE

The percentage of adjunct which can be used in the grist is limited by certain factors:

1. They cannot be used in amounts over which they dilute the flavour of the malt.
2. They cannot be used in amounts over which their own flavour becomes too dominant.
3. In the case of cereal adjuncts, they need sufficient enzymes from the malt to convert their starch to fermentable sugars, as they do not have their own or enough of their own enzymes to do so. This has been overcome to a certain extent by the use of supplementary commercial enzymes with the adjunct.
4. Malt used alone to produce a wort supplies sufficient nutritional requirements for yeast, but excessive use of adjuncts can dilute the level of some essential nutrients in a wort and stop the fermentation before all the sugar has been fermented.

BENEFITS

The main benefits to the brewer in the use of adjuncts are:

1. They can show economic savings.
2. In the case of copper adjuncts they can be used to increase the length or strength of the brew without expenditure on mashing plant. They then are known as wort extenders. This has been useful in recent times with the advent of high gravity brewing, in which brews are brewed at a higher gravity than required and then diluted after fermentation.
3. They are used to dilute the nitrogen content of a wort. The nitrogenous component of wort is potential haze forming material, so if it is reduced the chance of a haze-forming is reduced.

4. Syrups are now produced with fairly precise sugar compositions which are suitable for specific brewing needs. A wort, for example, can be made more fermentable by adding a 100% fermentable brewing syrup.
5. Some adjuncts can be used to improve the beer head.
6. Specific flavours can be added to the beer.

CEREAL ADJUNCTS

Cereals all contain starch in their endosperm which is the food store for the plant to grow on germination. The starch is in the form of granules which differ from one cereal to another and require different levels of severity in treatment in order to obtain the useful carbohydrates. Gelatinisation is the effect produced by the application of heat and moisture to the starch granules which disrupts the cell walls so that enzymes can act. Some cereal starches gelatinise at mashing temperature, so these adjuncts can be added directly to the mash. Others have to be brought to a higher temperature in another vessel in order to gelatinise them before they can be added to the mash. The cereals used as adjuncts include maize, rice, barley, wheat, oats, rye, triticale and sorghum.

Various preparations are used to suit the characteristics of the different cereals. Maize and rice cannot be used with the germ attached as it is oily and contains a lot of protein material. The endosperm is therefore detached from the rest of the grain before its use in brewing. This product is known as grits. Maize and rice gelatinise at a high temperature so have to be heat treated before they are added to the mash. This can be done by:

1. Cooking them in the brewery in a cereal cooker.
2. Flaking the endosperm during manufacture. In this process the grits are steamed or pressure cooked and then passed through heated flaking rolls at 84°C. The flakes are then dried so that they will keep without spoiling. The process was first used in the U.S.A. in the 1880s. A more modern method of heating the grain is by micronisation. In this process infra-red waves are used to heat the grains to 140°C before they are flaked.

Barley and wheat are gelatinised at the normal mash temperature, although pre-cooking is sometimes practised in order to get better utilization. Pre-cooking can cause an increase in viscosity which has to be balanced against any benefits. An alternative is to mill the grain, as the starch gelatinises more easily when the particles are smaller.

ONE OF THE FIRST FLAKING MACHINES

Wheat and barley may also be used in the form of flakes which are made in the same way as already described, using the whole grain. Care has to be taken in the manufacture of flaked barley if a viscous wort is to be avoided, but this can be overcome by spraying the grain with suitable enzymes during flaking.

Another form of wheat and barley which has become more popular in recent times is the torrified form. The torrification process was patented by Mr Fordred of Tottenham in 1882 though not immediately used for manufacturing the product. The first time the product is recorded as being manufactured is in 1886 when it was first used in the U.S.A. In the process the grain is rapidly heated to 260°C, when the sudden expansion of water in the grain puffs it up, sometimes bursting it to give a pop-corn-like appearance. The starch is gelatinised and also some colour and flavour develop. Traditionally this was done by passing the grain through a stream of red hot sand. The more modern process is to heat the grain with infra red heat as in the production of micronised flakes, but instead of flaking it, the grain is just cooled. Torrified wheat and barley are crushed with the malt in the grist when used.

MAIZE

This cereal is very widely used throughout the world. In Britain it has been used since 1880 when its use was permitted under the 'Free Mash Tun Act'. Although it is not a native crop it has always been a popular adjunct, the flaked form being the most popular. However, in recent times the amount of flakes used has been exceeded by the amount of grits used. It continues to be used for traditional reasons despite its high price. Maize has been particularly popular in the Lothian area of Scotland, as the barley in this part of Britain has a higher than average nitrogen content, which can be conveniently diluted by its use. Maize flour is not used as it has to be gelatinised like the grits and is not so conveniently handled.

France and Belgium who never had the benefit of colonial sugar use maize grits as adjuncts almost exclusively at the rate of 20% to 30% of the total grist.

In Africa, with the exception of native sorghum beers, maize is also the most popular adjunct.

In the U.S.A. maize and rice grits are the most popular adjuncts and used at the rate of between 35% and 40% of the total grist. Maize is the more popular of the two as it does not gum up the cereal cooker so much.

Maize gives beer a full flavour when used.

RICE

After the Free Mash Tun Act of 1880 which permitted the use of raw grain in Britain, rice was the cheapest adjunct available for brewing beer. It was used in the form of flour or grits and particularly liked because of its neutral flavour. In 1883 Beanes introduced a form known as Beanes Patent Grist which was more readily utilised. It was prepared by the action of dilute sulphuric acid on rice grits over two days, which reduced them to a friable condition after drying. Flaked rice was then introduced in Britain and it is this form which became most popular, being only second in popularity to flaked maize before World War II. However, the price of rice compared to other adjuncts does not make its use attractive in Britain today and it is only used in American beers brewed under licence.

In the U.S.A. rice grits are preferred to flakes and although a popular adjunct, its use is declining because it is more expensive than maize.

The rice used in brewing is a by-product of the edible rice industry where up to 30% of the rice produced is removed as dehusked or broken grains. This percentage is being reduced and causing the price of brewing rice to rise.

Rice gives beer a light clean flavour when used as an adjunct.

BARLEY

The use of this cereal as an adjunct has been most popular in Britain. Though raw barley was used as an illegal adjunct before the Free Mash Tun Act of 1880, there was no rush to use it once its use became legal. The use of sugar had been made legal in 1847 and it had established itself as a popular adjunct, unlikely to be displaced by raw barley. There was some initial interest in 1883, when Thomas Lovibond wrote a book called *Brewing with Raw Grain* which followed from a series of articles in the *Brewer's Guardian*. He indicated that brewers had been experimenting with the use of raw barley and other grain, but expressed surprise that there had not been more interest. The book gave theory and details of use for all types of raw grain using both conventional brewing plant and special converters. The use of raw grain in brewing, for a time, looked to be the way ahead and a number of these converters were patented and installed in breweries. The Billings system which allowed the use of 50% raw grain is one which was well advertised and is the machine shown in the illustration.

The use of raw grain, however, did not catch the imagination of the majority of British brewers who preferred to stay with their conventional plant. It was felt that savings on materials were not enough to justify capital expenditure on the new plant and to pay for the extra labour required to operate it. The savings made by using sugar and flaked cereals, which could be used with no extra cost on plant and labour,

RAW GRAIN BREWING.

Billings' Converter.

BY this process 50 per cent. Raw Grain can and is being used daily with the most satisfactory results in large London and many Provincial Breweries who have adopted this Machine and Process. The attention of Brewers is specially called to the Process, **the only one that has stood the test of time.** In proof of this two large London Breweries have had the Machine in daily use for some Two Years without a hitch in the working of any kind, and a repeat order has been sent for a Second Machine from one of these Breweries. Several large provincial Breweries have also adopted the process with most satisfactory results in all cases.

ADVANTAGES–

1.–Increased Extract, Brilliancy, Palate Fulness, Keeping Qualities, and absolute Uniformity of Resulting Beers.

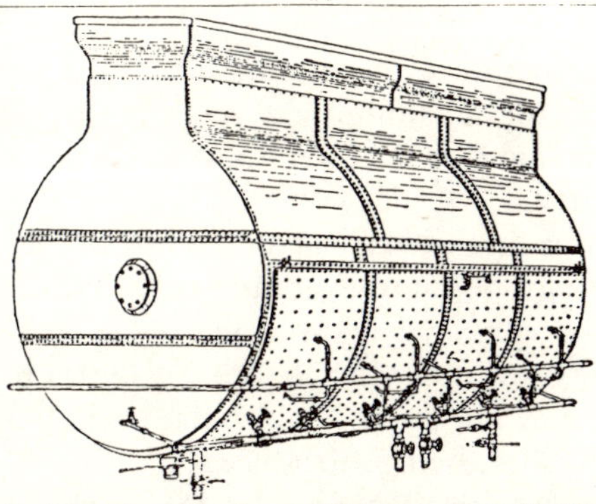

2.–Great Saving, Easy Working, Perfect Control of Temperatures.

3.–Yeast Crop Healthy and Vigorous. Beers Clarify Quickly and Fine Perfectly.

The saving is therefore ENORMOUS.

The ***Full Extract*** of both Raw Grain and Malt is secured, but raw, oily, and fatty flavours are Excluded. The Brewer has more ***control*** over the composition of the Wort than when dealing with fine Barley-malt in the usual way. The Machine can also be used under pressure if desired.

The ***Raw Grain*** and ***Malt*** are ***Converted in One Vessel,*** which afterwards takes the place of the Brewing Copper; a perfect ebullition is secured in it. The Copper and Hop Back may therefore be dispensed with. The beer may be aërated, hot or cold, in the same vessel.

THE CONVERSION COMPANY invites the most searching scrutiny of the whole Process—from start to finish—of the Beers produced, and of the Materials used. They will forward on application the reports upon the System of the leading English Scientific Authorities. They assert that this is the only Process extant for dealing successfully with large percentages of Raw Grain, and they will undertake to prove to Brewers (at Breweries where the system is in operation) that the claims they put forward are fully justified by actual practical working.

GUARANTEE—FLAVOUR and CHARACTER of Ales and Stouts UNCHANGED.

Pamphlets and Full Particulars sent on Application to

A. W. BILLINGS, Managing Director, The Conversion Company, Limited,

21, REGENT ST., WATERLOO PLACE,

LONDON, S.W.

Telegraphic Address, "BARLEY BIRD." **Telephone: 3940 Gerrard.**

were far more attractive. Many brewers who installed converters eventually had them removed. In the first Excise statistics where the usings of raw grain were separated from the usings of malt in 1897, raw grain represented only 0.4% of the total materials used.

Consideration was only again given in recent times to the use of raw barley after some research work in the 1960s. This followed some earlier work in the 1940s when the potential was seen for its use during the war. The relatively large increase in the price of sugars over the price of malt in the early seventies was the spur needed to develop brewing techniques so that raw barley could be used. The development of suitable plant and the use of enzymes as well as the low price now make its use an attractive proposition.

Barley used in small amounts can be milled with the malt. When larger amounts are used, which may be as high as 50%, the barley is milled separately from the malt in special mills, in order to obtain maximum extract. It is then added to the mash with enzyme preparations and a temperature programmed mash is used, which is again designed to obtain the maximum extract. Care is needed in setting up the system but it is said that once set up, it will give a wort similar to that from an all-malt grist.

Flaked barley was first used at the beginning of the twentieth century, but did not come into popular use until World War II when it was used as a substitute for other adjuncts which were not available. Its use continued in some breweries after the war. A better product is dehusked or pearl barley flakes but this has not been available in the U.K. since about 1970, shortly after being introduced.

Torrified barley was available in the 1880s but was little used. It is now more easily produced using micronisation and its use has been revived. At present flaked barley and torrified barley are less popular than raw barley. Torrified barley is less popular than torrified wheat as it does not give such a good extract.

WHEAT

Research into the use of raw wheat in the brewing of British beers was carried out in the early 1960s at the Brewing Industry Research Foundation around the same time as the work on raw barley. Wheat flour was chosen because it is readily available in Europe. The bran and germ are removed before milling and the nitrogen content can be reduced by air classification. Gluten is produced from the removed portion which helps keep the price competitive. The trend has been to use pelletised flour as it is easier to handle and is crushed with the malt.

Torrified wheat is an increasingly popular adjunct but the use of wheat flakes has died out in the U.K.

Wheat gives beer a dry flavour and improves the head.

In the twentieth century wheat beer, which has always been widely brew in Germany, is still being brewed and is even gaining in popularity. There are two main types: one brewed around Munich and the other around Berlin. Berliner Weisse (white) is a low gravity beer produced with about 25% wheat. Weizen (wheat) Suddeutsche is stronger and made with between one-third and two- thirds wheat.

The wheat beers which were produced in Austria, Switzerland, the Netherlands, Belgium and Denmark have also survived to some extent with the exception of the Netherlands. Belgium Lambic beer is perhaps the most famous of all wheat beers though it must be admitted that this is not entirely due to the fact that it is produced from wheat. It is also a spontaneously fermented beer which gives it other distinctive characteristics.

OATS

In the twentieth century flaked and malted oats have been used to give certain stouts their distinctive flavour. Malted oats are preferred and the stouts are known as oatmeal stouts. Oats have not been used in other beers as they give a poor extract, a high wort viscosity, poor fermentability and have a high lipid content.

SORGHUM

This is used as the main source of extract in some native African beers, but the astringent compounds in the pericarp of the grain give an unacceptable type of bitterness when it is used to brew European and American beers. The germ of the grain also has a high oil content. These characteristics led to a bad experience when it was used in the U.S.A. between 1943 and 1946 to replace corn adjuncts which were in short supply. The germ and husk are difficult to remove from the grain and were only partially removed from the sorghum used. More modern milling techniques now allow a pure grit and flour mixture to be produced. This has been used to produce an acceptable beer and can be used in the same way as corn and rice.

RYE

Rye is not widely grown in Europe so has it has rarely been used as an adjunct. It produces a wort with poor fermentability and viscosity. It is used with oats in the traditional Russian beer Kivas or Quass.

TRITICALE

This cereal is a cross between rye and wheat. It has shown some potential in trials producing a wort with a better fermentability than rye or oats, but it has a high nitrogen content.

ROOT STARCHES

Potato starch (farina) can be used in the mash tun as an adjunct as it gelatinises easily. The *Country Brewer's Gazette* published details of a process for using potatoes in the mash tun in 1878, but the idea was not taken up by British brewers as the idea of producing beer from potatoes was not acceptable to the public, even as it is today. Its attractive price has not so far tempted brewers to use it, except very occasionally.

Cassava (manioc) root has been used as an adjunct in the tropics in the countries in which it is produced, but not elsewhere. It has to be refined to remove the toxic substances which are also found in the root, but this is done as a matter of course whatever food it is used in.

BEANS AND PEAS

Beans are not generally considered acceptable adjuncts for brewing, although the soya bean because of its use in vegetarian and other food preparations, has gained some acceptability. Its use has been restricted mainly to very small levels in the U.S.A. It contains about 42% protein which is very high compared with cereals but little of it dissolves, so it can be used with low nitrogen malts in the mash tun. Flaked beans from which the oil has been removed during processing is the most convenient form in which to use it. Its advantage in use is that it gives improved fermentations and improved head retention to the beer produced.

Peas are now not used in brewing.

SUGAR-RICH ADJUNCTS

These have always been used in greater amounts than starch rich adjuncts in the U.K., particularly in the past when there was access to cheap Commonwealth sugar. They also have the attraction of being easy to use and as wort extenders offer greater production flexibility. At one time sugars were the cheapest adjuncts but now they are the most expensive. They are still widely used at much the same rate as they were in the past, as to many brewers their practical advantages outweigh their disadvantage in price. High gravity brewing, in which the wort is brewed at a higher gravity than required and then diluted with treated water, has had much to do with this. The problem with brewing a strong beer and retaining the usual brewing length is that the mash tun is sometimes too small to accommodate the extra malt required. The use of sugar as a wort extender in the copper is a convenient way of overcoming the problem. There is also undoubtedly a reluctance on the part of brewers to stop using sugars in their beers if they have been using them for many years as a change in the flavour could adversely affect sales.

Sugars can be classified as:

1. Cane or beet sugars.
2. Sugars derived from starch.

CANE AND BEET SUGARS

These were allowed for use on a permanent basis in Britain in 1847. Although refined cane sugar was not immediately popular it became so as the demand for sweeter, lighter-bodied beers increased towards the end of the nineteenth century. The taste for the acidic, vatted old ales and porter was in rapid decline.

Cane and beet sugars were the traditional sugars which used to dominate the market until relatively recent times, but they have been partially replaced by syrups produced by the hydrolysis of starch. The fall in usage started in the period of high inflation in the U.K. between 1973 and 1975. Their price more than tripled in the two-year period, whereas the price of sugars derived from maize starch only just about doubled. This diminished their use as a copper adjunct but they continued to be well used as priming sugars and for giving special flavours.

The only other countries which use much sugar for brewing are those which produce it themselves or are near to a country which does. Australia, for instance, has used sugar extensively for many years. It was used as refined white sugar until World War II when bagged raw sugar became the only source available. This was used until the late 1960s but now liquid sugar is used almost exclusively.

In the U.K. when cane sugar was first used it was either raw or crudely refined, sometimes producing disastrous results, but now all sugar undergoes refining. The degree of refining affects the flavours and colour left in the sugar. The darker more luscious flavoured sugars are used mainly to give beer an improved flavour whereas the white, bland, highly refined sugar is used solely to provide extract.

Invert sugar was the most popular type of sugar used for brewing between 1880 and 1960. It is manufactured from cane sugar by treating it with a small amount of sulphuric acid or hydrochloric acid, which catalytically breaks the sucrose molecule into glucose and fructose. After this has occurred the acid is neutralised and some further purifi-

MANBRÉ

PURE MALT SACCHARINE.

Extensively used by the Best Brewers in the Kingdom.

The fermentable Saccharine in Malt Worts will remove only about two-thirds of the nitrogenous matter it contains. The remainder is principally the cause of Beer turning sour; and the use of our Saccharine, in the proportion of one-third Saccharine to two-thirds Malt, will remove all remaining nitrogenous matter, and free the Beer from objectionable tastes.

It will increase the amount of alcohol and carbonic acid, and thereby augment the keeping property of the Beer, and make it specially adapted for export to hot climates.

It produces Beer greatly excelling in purity, in brilliancy of colour, and delicacy of flavour, the Beers brewed with Malt alone, or with Malt and Cane-Sugar.

As it is sent out in small pieces, the Brewer will find no difficulty in dissolving it either in the Hop or Underback, by letting the Worts run over it, and where space is limited it will enable the Brewing of one-third more Beer without increase of Plant.

INVERT SUGAR.

We manufacture also a very superior description of Invert Sugar or "Saccharum," which we guarantee to be made solely from Cane-Sugar. This article is made upon the most improved principles, and as none of the crystals or "pieces" are abstracted an absolutely pure product is ensured. We supply it as a liquid in 2 cwt. and 4 cwt. casks, or as a solid in 1 cwt. galvanized iron covered pails.

For Samples, Prices, Terms, and References, apply to

MANBRÉ SACCHARINE COMPANY, Limited,

SOLE MANUFACTURERS.

Works: Fulham Palace Road, Hammersmith, W.

Offices: 110, Cannon Street, E.C.

IMPORTANT!!!

256 lbs. of our products may be used as an equivalent for one Quarter of Malt, which places them in a very favourable position in respect of Excise Duty, as compared with Malt.

cation is carried out. The sugars produced are either used as a syrup or seeded with crystallised sugar and cast into a solid block. As the sugar has already been inverted, the yeast does not have to invert it during fermentation. It is believed that the yeast, because part of the work is done for it, will then be able to ferment the sugar faster whilst maintaining itself in good condition. In 1960, bulk liquid sugar which is transported in tankers started to become more popular than solid sugar because it is easier to handle. This type of sugar is mainly sucrose which is not inverted, so there has been a decrease in the amount of invert used for this reason. Though it only gained popularity in the 1960s it is of interest to note that bulk liquid sugar was first used in 1933.

Beet sugar cannot be used in any form other than a highly refined state and because of this its use in brewing was not even contemplated for many years. If not highly refined, the impurities of nitrogenous material affect the flavour of the beer giving it an unpleasant aroma. Early attempts to use it were a disaster and prejudiced its use for a long time afterwards.

GLUCOSE

The process of acid hydrolysis has been used since the early nineteenth century to make confectioners' glucose from starch. It is a mixture of glucose and intermediate sugars. Essential steps in the process include the breaking down of the starch with acid using heat and pressure to depress the boiling point, neutralisation of the acid and purification with charcoal. The longer the breakdown of the starch is allowed to take place, the more glucose there is produced and less of the intermediate sugar products. The resultant sugar mixture is cast into blocks after being seeded. This type of sugar is very hard so the blocks are broken up into chips which led to it being known as chip sugar or glucose chips. It can also be supplied as a syrup when blended with other materials.

Most glucose manufactured initially was for the confectionery industry, but what was suitable for them was not necessarily suitable for the brewing industry. In Britain, before about 1885, quite a lot of German potato glucose produced for the confectionary industry was used in brewing, but it was found that a better quality product for brewing was obtained from sago and maize. By the end of the nineteenth century 70% of British glucose used for brewing was being produced from sago. American glucose made with maize was also being imported.

British manufacturers started producing a type of sugar manufactured in the same way as glucose, but hydrolysis was not carried out so completely in order to produce something more attractive to the

"LION BRAND"

SACCHARINES.

"LION BRAND"

DEXTRIN-MALTOSE.

FULL PARTICULARS UPON APPLICATION TO

L. NORMAN & CO.,

3, EAST INDIA AVENUE,

LONDON, E.C.

brewer. Cornelius O'Sullivan and William Valentin, in 1874, had patented a process for producing from rice starch, a sugar which consisted of dextrins and maltose in the same proportions as found in wort. This glucose type of product was named Dextrin-maltose. Another similar product was Dextro-saccharum produced from sago by Garton, Hill & Co. Such products were known as saccharines rather than glucoses but should not be confused with the chemical sweetener with the same name which was prohibited for use in brewing.

Glucose gives beer a dry flavour but has the disadvantage that it has a very bitter taste produced by hydrolysis of some of the glucose during production.

WORT SYRUPS

Wort syrups were first used in the U.S.A. and Britain at the very end of the nineteenth century. They were produced by the hydrolysis of maize starch with acid as described in the production of glucose. Manufacturers aimed to match the maltose and dextrin content of wort by altering pressures and times of hydrolysis during production. The syrup was quite pure after overcoming initial problems in removing the oil from the grain. Although it was cheaper than invert sugar it had a dry flavour which did not make its use acceptable to all brewers. It became known as corn syrup.

In the U.S.A., around 1940, an improved product was manufactured using enzymes as well as acid for breaking down the starch. Hydrolysis which breaks down the starch occurred in two stages. In the first stage the starch was liquefied with dilute hydrochloric acid. In the second stage the dextrins produced by liquefaction were broken down to the required sugars by enzymic action. The product had a greater fermentability than the older types of syrups and more closely matched the sugar composition of wort. It has been further improved since then with enzymes replacing the use of hydrochloric acid in the first stage. Syrups are produced with different fermentabilities which suit specific brewing applications. In Europe wheat is cheaper than corn and a syrup is produced from gluten-free wheat starch.

MALT EXTRACTS

These are produced by evaporating the water from a normally produced wort in order to concentrate it. There are two types of malt extract: diastatic and non-diastatic. The diastatic type is made from a mash of a highly diastatic malt (one in which the enzyme level is very high). It is used as an adjunct when there is a desire to boost enzyme activity in the mash tun. It has tended to become obsolete in recent times as enzymes can now be added to the mash directly. The non-diastatic type is useful as a wort extender in large breweries, for use in small breweries which have no mashing plant or as a convenience for home brewers. In the later case it is also supplied ready hopped so it does not even have to be boiled.

Robert Thornton, in 1778, was the first person to produce a concentrated wort. It was solid and as brittle as glass as all the moisture had been evaporated out of it, using a double skinned steam boiler in order to prevent the wort charring. The idea of using it on naval ships had been suggested by Henry Pelham, Secretary of the Commissioners of Victualling to the Royal Navy. Two extracts were produced by Henry Jackson and tested in 1779: one was fermented extract and the other unfermented extract but both were hopped. After an initial declaration of success there was a change of view and the idea was abandoned for producing beer on board ship. The extract was then only issued as a medicine which was supposed to prevent scurvy.

In the 1870s, Professor Reitch in Bohemia also produced a crude malt extract which was in a solid, glass-like state. He called it Zelithoid. The first malt extract which was of practical use to the brewer was not produced until early in this century in Essex. It was in the form of a thick syrup.

In the 1960s green malt extract was developed which was also known as liquid malt. The aim was to by-pass the expensive kilning process of malting by completely replacing malt with green malt extract. However, this did not produce a satisfactorily flavoured beer and the highest percentage that could be used was 50% replacement. Interest in the product had waned by 1975.

By some definitions malt extracts are not considered adjuncts as they are malt products, but as they are used in the same way as adjuncts it is useful to examine them as well.

BARLEY SYRUP

This is a syrup which is made from raw barley and has a similar composition to the wort produced from malted barley. It has a mild but acceptable flavour, free from the taint of raw grain. It is made in a mashing type of process using specially milled barley, enzymes and a little malt. It was developed in the 1960s but like green malt extract its use was not taken up by brewers. They probably preferred to be in control of their own wort production using raw barley and enzymes in the mash tun.

This has become the more popular approach to the use of raw barley by the brewer.

PRIMING SUGARS

Various sugars are used for this purpose and are added to beer after the primary fermentation. They carry out a dual role:

1. To give sweetness and flavour.
2. To provide fermentable sugar for a secondary fermentation.

They were first used at the end of the nineteenth century in weaker running beers, which did not contain as much residual sugars as the stronger beers. The priming sugars were fermented rapidly and enabled the beers to mature quickly. Initially the amount of priming sugar solution added was limited to 1.5% of the volume of the beer and had to have a gravity no higher than 1150. There was some feeling at the time that when priming sugars were added to a beer, they gave the impression that the beer was much stronger than the gravity to which the primings had raised it and that because of this revenue could be lost. There is now no restriction on the amount of primings added.

Every manufacturer of priming sugars has his own blends of various types of sugar including cane, invert, glucose, starch syrup and lactose. A standard mixture which does not crystallise out is 55% invert sugar and 45% cane sugar. Fructose is the sweetest of sugars and a syrup containing a high percentage can now be produced from glucose syrup using the enzyme glucose isomerase. Lactose is the sugar found in milk from which it is extracted and is important as a priming sugar as it is sweet but unfermentable. If a lot of sweetness is required, as in some stouts for example, it is not practical to obtain it by adding a large amount of fermentable sugar. If this was done problems with biological stability would result as it would provide a nutrient source for the development of spoilage micro-organisms.

This is an appropriate point to mention Milk Stout, which derives its name from the fact that it contains lactose, or milk sugar, not milk. Mackeson acquired the letters patent in 1908 for the brewing of Milk Stout and granted licences to other brewers for its manufacture. The grist contained 9 lb/brl of lactose. A brewery using the name without a licence, in 1912, was fined under the Merchandise Marks Act 1887. The name under present law could not be used as it is regarded as a misleading description.

SACCHARINE

Saccharine, the well-known artificial sweetener (also known as synthetic or non-nutritive sweeteners) was synthesised in the U.S.A. in 1879. The advantages of its use in brewing, with the popularity of sweeter beers, was immediately seen by brewers but its potential threat to the revenue was just as quickly seen by the Excise. Under the Inland Revenue Act of 1888, by giving notice in the *London Gazette*, the Inland Revenue could prohibit the use of substances of a 'noxious or detrimental nature' or which 'may affect prejudicially the interests of the revenue'. Saccharine was immediately prohibited. In general any substance which in the opinion of the Commissioners gives the impression of enhanced original gravity without actually producing a measured increase in gravity is still prohibited. As well as saccharine and its salts, glycyrrhizic acid and its salts are now also prohibited.

Some countries such as Belgium permit the use of saccharine and legislation brought in between 1960 and 1970 allowed up to 150 ppm in some beers.

COLOURING

Colouring is the name given to substances which are used to increase the intensity of the natural colour of beer. When colouring is added to beer it becomes more brown and if sufficient colour is added it becomes black.

DEVELOPMENT OF THE NEED FOR COLOUR

The necessity to add colour to beer did not occur much before the beginning of the nineteenth century. Prior to this time the colour of the beer was produced entirely from the natural colouring substances in malt. Early malting methods mostly relied on the use of wood for kilning, producing a malt which by today's standards would be classified as brown. This in turn produced a dark coloured beer. There is a tendency for beers to darken in the cask when they are left to mature, as was the general practice in the eighteenth century. Also the more malt used in a brew, the darker is its colour. Thus the darkness of the beer at that time was partly an indication of maturity and partly an indication of strength, but there are no indications that beer was coloured in order give a false impression to the drinker.

In the seventeenth century coke was first produced in Derbyshire and was used for kilning malt, allowing the production of pale malts and pale ales. It was not until the end of the eighteenth century that the work of Richardson on the scientific use of the saccharometer, revealed that the pale malts gave considerably better extracts than the brown malts which were in more widespread use. In one case, in 1781, he showed that a pale malt and a brown malt made from the same barley produced extracts of 82 and 78 pounds per quarter respectively. Brown malt made at Ware was a type of malt much in favour by the London porter brewers and in the same year it produced an extract of only 56 pounds per quarter. This particular kind of brown malt was known as blown malt as it was blown up in size by the sudden application of violent heat, as in the production of popcorn. The increased size was visually impressive to the prospective buyer but the poor extracts which it produced were not apparent. They only became so when the saccharometer came into common usage.

By the end of the century prices of raw materials had risen so much that the porter brewers could not ignore the benefits of using pale malt. However, the problem in using it was that the dilution of the colour of the beer produced from it restricted the amount which could be used. There was thus an incentive to find some substance which could be used to colour-up beer.

EARLY COLOURING SUBSTANCES

An isolated example of the use of colouring in the eighteenth century was in a recipe for Brunswick Mum. Walnut, madder and red saunders were used to increase the intensity of colour. However, this was a special type of beer and the cost of such ingredients would have made them prohibitive for use in ordinary beers even if permitted.

The common form of colouring first used in porter was prepared from molasses. This was boiled until it became dark, thick and bitter. At this point it was set on fire for five to six minutes and then after extinguishing, it was diluted back to the original consistency.

A better product was *essentia bina* which was prepared from muscavados or raw sugar. This had to be made in a special iron boiler fitted in brickwork with a grate and a flue. An internal scraper was also fitted inside the boiler which could be rotated in order to keep the sugar in motion so that it did not burn on the bottom. The sugar was boiled until the point was reached at which a sample taken out of the boiler quickly set on cooling. Again it was set on fire for a short time and extinguished by placing a cover over the boiler. Lime water was then cautiously added a little at a time with the stirrer kept in motion until the mixture reached the consistency of treacle. Lime water could not be used if heading was to be added to the beer as the two reacted. Water, wort or a batter of brown malt meal were used as substitutes. The rate of usage of

PLUNKETT BROTHERS'

PATENT MALTS,

PATENT
CHOCOLATE MALT,
BROWN MALT,
CANDIED MALT,
FOR
PORTER, STOUT,
AND
EXPORT.

CANDIED MALT,
GOLDEN MALT,
FOR
FLAVOURING
MILD ALE.

BELLE VUE MALTINGS, DUBLIN.

JAMES'S GATE BREWERY, DUBLIN, 25*th February*, 1873.

DEAR MRS. PLUNKETT,

We have the pleasure of stating that we are purchasing considerable quantities of your Patent Brown Malt and find it of very good quality, we would also say that our connection with your Firm and that of your husband the late MR. RANDAL PLUNKETT, and his father, extends over fifty years, during which we have had large and satisfactory transactions.

ARTHUR GUINNESS, SONS & CO.

Mrs. Eliza Plunkett,
Belle Vue.

ESTABLISHED 1819.

TRADE MARK.

essentia bina in porter was about 2 lb per barrel, either in the copper or in the fermenting vessel.

In 1802 Matthew Wood patented a colouring matter produced by concentrating wort until it was a paste-like consistency. However, this was banned as an adulterant although it was produced from malt. This was perhaps because the company in which Wood was a partner, Wood Wiganth & Co, were wholesale druggists who were under suspicion for supplying illegal drugs to brewers. In 1809, De Roch patented an idea for roasting malt skins. The scheme was to add them to the mash or to make an infusion in water. It is not known whether the idea was put into practice but if it was, it was not widely used. Burnt flour and liquorice were other substances which were tried but were not greatly popular.

In 1811, the need for colouring was officially recognised and colouring made from muscavados was allowed to be used in porter by Act of Parliament, but was repealed in 1816. The following year Daniel Wheeler patented, 'A new and improved method of drying and preparation of malt'. This very dark roasted malt produced under Wheeler's patent was known as Patent Malt and was quickly adopted by breweries. Whitbread began using it straight away and Barclay was using it by 1820. Two cousins named Plunkett set up their roasting houses in Dublin in 1819 and became suppliers to Guinness, the well known brewers of porter and stout.

COLOURING IN PORTER

The use of Patent Malt in porter allowed brewers to make it as dark as they wished. It became fashionable by the 1840s to make porter black, rather than the brown which it had formerly been. This was lamented by Black writing in 1844. He regarded it as a move to disguise any cloudiness in the beer and disliked the name Blackstrap which was used 'in common parlance' for porter. In fact it was probably a move by brewers to give it a distinctive appearance, making it different from the unmatured porter known as mild which was becoming more popular.

The change from brown malt, to pale malt used with colouring also changed the flavour of porter. Porter used to have an empyreumatic (burnt organic material) flavour but with the use of pale malt and colouring this flavour was absent and the taste came more into line with the cleaner flavour of ale. One would have expected that this would have made it more acceptable but this was not reflected in its declining popularity.

LATER DEVELOPMENTS

During the nineteenth century the production of colouring substances by the heating of sugar developed. The well known sugar manufacturer, Alexandre Manbré, took out patents in 1858 and 1859. The later one included a number of different substances from which it was proposed to extract the sugar and covert it to colouring material. The list included: 'rice, maize, carrots, maple, bananas, pompions, chestnuts, mangel wurtzells, sorgho, turnips and Jerusalem artichokes'. Eventually with continuing development the product which we recognise today as caramel was produced.

CARAMEL

The manufacture of caramel involves some very complex chemical reactions which are not completely understood, so there is little scientific information available on the manufacturing process. It has remained an art rather than a science in the hands of manufacturers who have used traditional and confidentially held methods.

There are three types of caramel:

1. Electropositive caramel made with ammonia, which accounts for 75% of total caramel production.
2. Electonegative caramel made with ammonium bisulphite, which accounts for 20% of total caramel production.
3. Spirit caramel which is made to dissolve in alcoholic and acetic solutions and is made without ammonia.

The electropositive caramel is used in beer as it does not interfere with the action of finings which are also electropositive.

The sugar from which caramel is now made, is glucose syrup, but over the years a variety of substances have been used. Before World War II, glucose made to a high standard from maize was used, but during the war sucrose, invert sugar and glucose made from sago, tapioca and potato were used. Sucrose continued to be used until the price became prohibitive. Now glucose made from maize starch is used again. Sucrose produces caramel with the best taste but it is not so stable.

The general principle of manufacture is as follows: Syrup and 0·880 ammonia are mixed and left for a week at ambient temperature. More ammonia is added and the mixture is held at 90°C overnight, whilst being stirred. Next day the mixture is heated driving off excess ammonia. An exothermic reaction occurs and the mixture boils at 120°C. The reaction subsides after three hours, the colour increases and the acidity increases. There is some further boiling to obtain maximum colour without making the caramel too viscous. The mixture is then cooled and water is added.

The use of caramel in beer has been examined by the Food Advisory Committee to the Ministry of Agriculture, Food and Fisheries who published their findings in an Interim Report (1979) and a Final report (1987) on the Review of Colouring Matter in Food Regulations 1973. The Interim report was particularly concerned over the use of caramels in food as they account for 98% of all the colouring used. They made several recommendations which were particularly relevant to brewers, as caramel is the only colouring which can be added to beer according to the Colouring in Food Regulations 1973. They were concerned that the product could not be chemically defined and that any results from toxological tests could not be extrapolated from one product to another. They therefore asked for the number of products, which numbered over a hundred, to be reduced and data to be produced on the reduced number. The British Caramel Manufacturers' Association (BCMA) had proposed six groups for caramels produced in catalysed reactions on which information would be produced.

After the Interim Report the International Technical Caramel Association (ITCA) which represents caramel manufacturers throughout the world, produced new specifications and data on characteristics which overrode those produced by the BCMA. The FAC were not particularly happy with these specifications which they considered to be too general and broad based to be of much value. They asked for answers to certain questions in order to make the proposals of the ITCA more acceptable.

Another committee of the MAFF, the Committee on the Toxicity of Chemicals in Food, Consumer Products and the Environment (COT) also looked into the use of caramels and their reports were published with the FAC reports. They too were disappointed with the ITCA proposals and found it difficult to relate the toxological data provided to the specifications. They also asked for further information on the lymphocytopenia observed after feeding caramel to experimental animals. This relates mainly to the ammonia caramels which are used in beer. Large amounts of this type of caramel can be consumed by some people who drink certain types of dark beer and could exceed the Acceptable Daily Intake (ADI) level set by the EEC Scientific Committee for Food (SCF).

On these grounds the FAC recommended significant reductions in the levels of caramels in food. They also advised that they would recommend further reductions if adequately reassuring information was not produced when the situation was reviewed in 1989.

A table in Annex III of the FAC Final Report shows the levels of caramel added to beer. The FAC recommended maximum is indicated as 5,000 mg/kg and was exceeded in some stouts, brown ales, milds and strong ales examined. Of the twenty-seven stouts twenty-four exceeded the set limit, of the seventy-nine milds it was four, of the forty-one brown ales it was six and of the thirty-nine strong ales it was only two. The lighter beers were all under the limit.

Considerable work has been carried out on the problem of lymphocytopenia caused by ammonia caramel. Lymphocytopenia is seen as a reduction in the lymphocyte count in the blood and an increase in the neutrophil count. Rats were used for experiments and symptoms appeared within days of starting to feed caramel. A component of caramel which has been shown to produce these effects is 2-acetyl-4-tetrahydroxybutyl-imidazole (THI), but indications are that it is not the only possible substance causing the effect. It has been recommended that as a precautionary measure THI levels should be below 25 ppm in caramel although there is no evidence so far of detrimental effects on humans. The EEC will implement the limit of 25 ppm THI in the eighth amendment to their Directive on Food Colour and the limit will be reduced to 10 ppm after three years. Since the levels of THI in brewing caramels have been reduced the Brewers Society has made representations to have the maximum permitted levels of caramel in beer increased.

A related substance which had been detected earlier in some ammonia produced caramels and reported at various times, between 1973 and 1977, was 4-methyl imidazole. This was found to have toxic effects on rabbits, mice and chickens. The World Health Organisation (WHO) recommended a limit of 200 mg/kg in caramels at a colour of 20,000 EBC units. Samples of ammonia caramel as used in beer were analysed at the Brewing Research Foundation in 1977 and gave results between 57 and 283 mg/kg. These figures were below the limit when they were adjusted to allow for the lower colouring value of the standard. Other brewing materials which contribute colour to beer were also analysed including: roasted barley, chocolate malt, crystal malt, and farbe bier. Only in one sample of roasted barley was any detectable and this was at a concentration of 1 mg/kg.

In June 1988, The British Caramel Manufacturers' Association published a leaflet: *Caramel – An Update*. In it they stated that they had made a detailed response to the FAC in a 500 page document, which satisfactorily answers all the questions raised in the FAC report. The leaflet also reported that the ITCA had produced a classification for caramel which is accepted by the FAO/WHO and the SCF in its draft eighth amendment to the EEC Colours Directive.

The classification is:

CLASS I	Caramel prepared by the controlled heat treatment of carbohydrates with or without the presence of alkali or acid.
CLASS II	Caramel prepared by the contolled heat treatment of carbohydrates with sulphite containing compounds.
CLASS III	Caramel prepared by the controlled heat treatment of carbohydrates with ammonia.
CLASS IV	Caramel prepared by the controlled heat treatment of carbohydrates with ammonium and sulphite containing compounds.

The caramels used by brewers are in CLASS III.

The BCMF indicated that they supported the classification and urged that British Authorities also adopted it.

Concerning lymphocytopenia observed in experimental rats, they reported this to be a readily reversible effect resulting from levels of THI above 25 ppm, but only where the diet also had a low level of Vitamin B6. All members of the BCMA were stated to have available, caramels of CLASS III, in which the THI level is below 25 ppm.

OTHER COLOURINGS

Other dyes are permitted for use in beer in the U.S.A. and include: Yellow No. 5, Red No. 40 and Blue No. 1. Yellow No. 5 can cause allergic reactions such as sneezing and asthma. There was controversy over Red No. 40 causing cancer but the FDA concluded that it did not. Why the Americans require these colours in beer is puzzling!

ADJUNCTS AS ADULTERANTS

Adjuncts clearly cannot be described as additives as defined in the introduction to this book. They are regarded as main ingredients by the British brewing industry. Some people, however, choose to regard them as adulterants. They present no health threat to the customer as they are no less wholesome than malt. The only other way in which they can be regarded as adulterants is if they are used to cheat the customer, that is, if by their use the customer is failing to get value for his money. The way in which the customer does not get value for money is if he gets something inferior to what he expects. In the absence of an official definition of beer which indicates what ingredients can be expected and in what amounts, the consumer has no justification in complaining that he is not getting what he expects. There may be a case for legally defining the amount of meat and the amount of rusk in a sausage, because the rusk is inferior to the meat in nutritive value and in the expectations of the customer, but in the case of beer it is difficult to argue that the extract obtained from adjuncts is in any way inferior to the extract obtained from malt.

The consumer may know that the brewer is saving money by using certain adjuncts but this does not necessarily mean that they are inferior sources of extract. The brewer, by selecting suitable cheaper adjuncts, is merely using the natural variations in the price of raw materials to his advantage, which is eventually passed on to the customer. Without some freedom the brewer could not be competitive and without competition the price of beer could rise against the interests of the consumer. It could be argued that the consumer is not getting what he expects when adjuncts are used because he is cheated of the full degree of flavour which would be obtained from the malt. However, in Britain the brewer uses adjuncts not only for economic reasons but because they are essential in providing the variety of flavours which are much appreciated by consumers.

Despite these arguments consumers may feel the need for some protection. Although they may concede that there is a case for the use of adjuncts they may see no reason why they should not know what ingredients are being used to brew the beer which they are drinking. A total ban on the use of adjuncts as under the German Reinheitsgebot would ensure that the consumer is never misled. A compromise would be to produce a statutory definition of beer under which permitted adjuncts are made known, with restrictions on the amounts which could be used. This could be developed further by making it a legal requirement for ingredients to be listed on labels, as they are for other foods. In the case of draught beers the details would have to be displayed on pump clips, dispense fonts or on notices in the bar. The idea is not new and was a feature of the Pure Beer Bill with failed to get through Parliament in 1896. Details are given in Chapter 9. The principle of labelling ingredients is opposed by the Brewers' Society on the grounds of practical difficulties but is strongly supported by the Campaign for Real Ale (CAMRA) and other food consumer organisations.

CHAPTER 8

UNUSUAL BEERS

HERB AND SPICED BEERS

Various herbs and spices were used to flavour beer before the introduction of hops, but there were certain beers which were particularly noted for the spices or herbs used in them. They were known as spiced or herb beers, or if the herbs used in them were of medicinal value, they were called medicinal or tonic beers.

SPICED BEERS

In mediaeval times, the use of spices in food was extremely popular with the people who could afford them, as much of the food was rich in starch which lacked flavour. The average diet also lacked the variety which is available to us in the twentieth century. Beer was no exception and spiced beer was a luxury which was sold at twice the price of common ale.

In Britain perhaps the most famous of spiced ales was Welsh ale, more specifically called Bragawd or Bragot. It was the most luxurious of beers and is recorded as having been paid as a tribute to a King by a free township. It was mentioned in the laws of King Ine of Wessex, between A.D. 690 and A.D. 693 and as land rent to the Church at Worcester in the time of Offa. A fourteenth century manuscript gives the following recipe:

> Take 10 gallons of ale, 3 potell of fyne wort, and 3 quartis of hony and put thereto canell [cinnamon] oz: 4, pepper short or long oz: 4, galingale oz: 1, and clowys [cloves] oz: 1, and gingiver oz: 2.

Spiced ales were also brewed on the Continent. The Church councils of Worms (A.D. 864) and Treves (A.D. 895) devoted part of their deliberations to spiced beer. They decided that church penitents were to be allowed to drink it only on Sundays. Common ale had to be drunk on other days.

Nutmeg was a popular spice as the following lines of verse of the time confirm.

> And notemuge to put in ale,
> Whether it be moist or stale,

and

> What gave thee that jolly, red nose:
> nutmegs and cloves!

HERB BEERS

Many different herbs have been used to flavour beer from the earliest times and the beers produced from them were known by the name of the herb used e.g., gale beer mentioned in Chapter 1. Despite the success of hops other herbs continued to be used in Britain, as we are reminded by Thomas Tryon in his*New Art of Brewing* written in 1691:

> Here give me leave to tell you that there are a great number of brave herbs and vegitations that will do the business in brewing as well as hops and for many constitutions much better; for 'tis custom more than real virtues that renders hops of general use and esteem; they are an excellent herb and would be much better, if they were or could be dried in the sun. Penny Royal and balsam are noble herbs and of excellent use in beer and ale; they naturally raise and cheer the drooping spirits, and open and cleanse the passages after a friendly way, and with a mild operation. And also they add great strength and frequency, and make brave well tasted drink, good to prevent and cure all, or most of those diseases which the wise ancients have appropriated that herb unto. The like is to be understood of mint, tansie , wormwood, broom, carduus, centaury, eye bright, betony, sage, dandelion and good hay; also many others according to their natures and qualities, and for those diseases to which they are respectively appointed.

Herb beers were more popular in country areas than in London and more in the north than the south. Most herbs continued to be used because of their medicinal or tonic values more than for their flavouring properties. These beers are described later in this chapter.

The herbs used for flavouring were predominantly bitter herbs which balance the natural sweetness from the malt sugars as do hops. Broom and wormwood were the two predominant bitter herbs used in Britain in the seventeenth and eighteenth centuries. Centaury, carduus (holy thistle) and dandelion were other bitter herbs used. More fragrant or spicy flavours were obtained from herbs such as iris, tansy, juniper, mint and pennyroyal. A complete list of herbs used in brewing, with their details is given in Appendix 1.

HEATHER BEER

This has a very long history and legend states that the Romans made it in Northumberland during their British occupation. Other sources state that the knowledge of brewing it was brought to Ireland and Scotland by the Vikings. In Ireland it was known as beoir-lochlannach, which in English is translated as 'strong at sea', a title by which the Vikings were known.

Scotland has a particularly old tradition of making heather beer which is believed to have stopped only at the beginning of the nineteenth century. Whether the beer was made entirely of heather bells or included malt and hops or just sugar is not clear, as the information varies from one source to another.

An interesting story is quoted by Bickerdyke:

> The tradition in Caithness runs that three Picts – an old blind man and his two sons – survived the rest of their race; that these alone of all mankind possessed the secret of making heather ale; that they guarded their secret with jealous care, and that they were in consequence much persecuted by their conquerors. At last the old Pict, in answer to the frequent importunities of his persecutors, promised to tell the secret, on condition that his two sons should be put to death. This was done, but the task was as far from accomplishment as ever, and nothing could be got from the old man but the truly Delphic words which were handed down in the couplet:
>
> Search Brockwin well out and well in,
> And barm for heather crop you'll find within.
>
> The secret died with him.

In 1985 archaeologists excavating a Neolithic settlement dating back 4,000 years at Kinloch on the island of Rhum, discovered some small drinking vessels on the inside of which was a crust. On analysis of the crust pollen spores of oats, barley, heather, meadowsweet and royal fern amongst other herbs was discovered. These are believed to be the base of a fermented beverage which was reproduced by researchers with some success.

Heather was also used in England in gale ale, a beer of Nordic origins, which was brewed in Yorkshire. Evidence of this is supported by information gathered from an old collection of vocabularies some of which date back to the fifteenth century. In two of them are words pertaining to the brewhouse, in which are found together, the two Latin words Pruera and Mercia. These are both translated into the English of the time as 'lingge' with a note that they were used for brooms. It was believed by Arnold that their use was wrongly assigned and that they were in fact ingredients. Pruera, he believed, was a corruption of Bruera which is ling-heather or ling and that Mercia was a corruption of Myrica or Myrica gale.

Heather was used in other places in England. Dr Plot who wrote a History of Staffordshire in 1686 mentioned of Shenstone, that 'they frequently used *Erica vulgaris* [*Calluna vulgaris*], heath or ling, instead of hopps to preserve their beer, which gave it no ill taste'.

MUMM

A mixture of herbs, rather than spices, was used in another famous beer called Brunswick Mumm, which originated in mediaeval times. It was said to have been named after its inventor Christian Mumme and first brewed in 1492. However, it was mentioned in earlier literature in 1350 and 1390, so this explanation cannot be true. An alternative explanation as to the origin of its name is that it came from the German verb mummeln (M.E. mumble), with various explanations as to the connection with the beer which stretch the imagination.

The beer was thick, strong and syrupy. A recipe, well quoted both in England and Germany, is believed to have come from a pamphlet of 1684 which was communicated to the British general, Monk.

> To make a vessel of 63 gallons, we are instructed that the water must first be boiled to the consumption of a third part, then let it be brewed according to the art with seven bushels of wheat malt, one bushel of oat malt, and one bushel of ground beans. When the mixture begins to work, the following ingredients are added: three pounds of the inner rind of the fir; one pound each of the tops of the fir and the birch; three handfuls of Cardus Benedictus, dried; two handfuls of flowers of Rosa solis; of burnet, betony, marjoram, avens, pennyroyal, flowers of elder, and wild thyme, one handful and a half of each; three ounces of bruised seeds of cardamum; and one ounce of bruised bayberries. Subsequently ten new-laid eggs, not cracked or broken, are to be put into the hogshead, which is then to be stopped close, and not tapped for two years, a sea voyage greatly improving the drink

Hacker writing in 1911, quotes an old source who has the following to say on the subject:

> Our English brewers use cardamum, ginger and sassafras, which serve instead of the inner rind of firr; also walnut rinds, madder, red saunders, and ellecampane. Some make it of strong beer and spruce beer; and where it is designed chiefly for its physical virtues, some add watercresses, brooke-lime, and wild parsley, with six handfuls of horse-radish rasped into every hogshead.

This obviously caused some confusion as the anonymous author of the

London and Country Brewer quotes a recipe in the eighteenth century which is a mixture of the two.

The beer was exported to England in the seventeenth century and became quite popular. Pepys wrote in his diary of 1664 that he went 'with Mr Norbury, near hand to the Fleece a mum-house in Leadenhall and drank there mum'. Pope (1688–1744) wrote:

> The clamorous crowd is hushed with mugs of mum
> till all, turned equal, send a general hum.

The beer, as implied earlier, was imitated and brewed in England to some extent. It was also imitated in Hamburg, as Hamburg mumme is recorded as being imported into Britain in the early eighteenth century.

PURL

Another beer, containing a mixture of several herbs, which was produced at the beginning of the eighteenth century was Purl, sometimes called Covent Garden Purl. It was initially a purging ale, made with senna, coriander, rosemary, sage and much Roman wormwood. The senna was dropped and a later recipe of 1802 included the following: Roman wormwood, gentian root, *calamas aromaticus*, horse radish, dried orange peel, juniper berries, seeds of Seville oranges and galingale root. A recipe from the previous century also included snake root. These herbs were immersed in a cask of beer for several months, usually from November until the onset of winter the following year. It was a winter drink and mostly drunk hot. It was common practice by some licensees to use waste beer, as the following note appears in the *London and Country Brewer*:

> This victualler is of opinion that there are scarce six in twenty of his fraternity in Town, who do not make their Purl only with their refuse or waste drink, such as they receive in their tap-tubs, by throwing into it no other bitter, but a parcel of common weedy wormwood; which compound one would think, more fit for a puke, than a grateful, cordial, stomachic beer.

OTHER EXOTICALLY FLAVOURED BEERS

The seventeenth century was a period when some exotically flavoured beers were produced. Like purl they were often made by suspending a mixture of the herbs and spices used in a linen bag in the beer. This was usually only for a few weeks but could be for longer. The following recipes for China ale, Apricot ale, Egg ale, Cowslip ale, Elderberry ale and Blackberry ale are all found in the *London and Country Brewer*.

The recipes would have been intended for licensees or private domestic brewing.

> [China ale] To six gallons of ale, take a quarter of a pound or more of china root thin sliced, and a quarter of a pound of coriander seed bruised; hang these in a tiffany or coarse linnen bag in the vessel, till it has done working, and let it stand fourteen days before you bottle it; though the common sort vending about town, is nothing more (at best) than ten shilling beer, put up in small stone bottles, with a little spice, lemon peel, and raisins or sugar.

> [Apricot ale] Take, to every gallon of ale, one ounce and a half of wild carrot seed bruised a little, and hang them in a linnen bag in your barrel, till it is ready to drink, which will be in three weeks; then bottle it with a little sugar in every bottle.

> [Egg ale] Take, to twelve gallons of strong ale, eight pounds of lean beef which must be cut into little bits and half stewed with a little water; and when it is cold, let the gravy be put into the vessel of ale, the fat being blown off; then let the beef with twelve eggs, their shells being only bruised, but the films not broken, a pound of raisins of the sun stoned, two nutmegs, a little mace and ginger, and two oranges cut round, be put into a linnen bag, and hang it in the barrel before it is done working; put in also two quarts of Malaga Sack, and stop it up; let it stand three weeks; then bottle it and into every bottle put a clove and a lump of sugar.

> [Cowslip ale] Take, to a barrel of ale, a bushel of flowers of cowslips pick'd out of the husks, and put them into your ale, when it hath done working, loose in the barrel without bruising, and let it stand a fortnight before you bottle it, and, when you bottle it, put a lump of sugar in each bottle.

> [Blackberry ale] Take two bushels of malt, and make it into strong ale, allowing a quarter of a pound of hops to it; when the wort is cold enough put into your vessel a little yeast and the juice of three quarters of a peck of blackberries full ripe, and ferment them all together; when it has worked sufficiently, stop it up close, and at six weeks end you may bottle it, and in a fortnight after, it will be fit to drink.

> [Elderberry ale] Take a hogshead of the first and strongest wort, and boil in the same one bushel of pick'd elderberries full ripe; strain off, and, when cold work the liquor in the hogshead, and not in the open tun or tub; and after it has lain in the cask about a year, bottle it, and it will be a most rich drink, that they call Ebulum.

ALES CONTAINING MEAT

One of the best known ales which was made with meat was cock ale. There are numerous recipes dating from the eighteenth century, but essentially it was made by putting a cockerel in a cask of ale with other ingredients such as raisins, mace, rosemary, cloves, cayenne, dates and honey. The cock was gutted, drawn and flayed then boiled and crushed. The cock and other ingredients were then put into a canvas bag and immersed in the ale for a week or ten days, after which it was bottled. The cock was sometimes soaked in Madeira or a similar wine before use.

Meat was also use in Egg ale as described above.

A beer which had an almost legendary reputation is Morocco. It was produced at Levens in the county which used to be called Westmoorland, the ancient seat of the Howard family. The recipe was believed to have been brought from abroad in the times of the Crusades. It was a very strong beer and never tapped under twenty years old. The recipe was buried during the Civil War and only came to light some time later, but was kept secret. However, it is believed that meat of some sort was an important ingredient.

MEDICINAL AND TONIC BEERS

Medicinal or tonic beers were another type of beer in which herbs played an important role. In these their medicinal properties were more important than their flavouring properties. Different herbs are encountered.

In Anglo Saxon times a recommendation for lung disease was for a man to:

> . . . withhold himself earnestly from sweetened ale, to drink clear ale, and in the wort of the clear ale, boil young oak rind and drink. Fever patients are recommended to drink during a period of thirty days an infusion of clear ale and wormwood, githrife, betony, bishop-wort, marrubium, fen mint, rosemary and other herbs.

and for trifling ailments such as hiccups for example it was recommended to:

> Take the root of jarrow, pound it, and put it into good beer, and give it to the patient to sup lukewarm.

There are certain herbs which for centuries have been known for their expectorant properties. These have been used in beers given to people to fight off the symptoms of coughs and colds. They include hyssop, St John's wort, elecampane and coltsfoot.

Rosemary, which acts as a tonic and has a camphoric aroma, has been added to beer for medicinal purposes from mediaeval times until the eighteenth century.

Betony ale was recorded in the nineteenth century as being an ancient ale of note. Though it is used as a flavouring the herb did have the reputation of being a 'cure-all' and for being the best herb for curing headaches. Betony ale is therefore most likely to have been a medicinal ale.

Nettle beer had the reputation of being good for rheumatic and gout pains. A twentieth century recipe is as follows:

Water	2 gallons
Nettle tops	2 gallons
Dandelions	4 handfuls
Clivers	4 handfuls
Bruised ginger	2 oz
Brown sugar	1 lb

Boil for forty minutes, strain , cool and ferment with yeast as in the production of ginger beer

A much earlier sixteenth century recipe for nettle beer was offered as a cure for black jaundice.

In the seventeenth century tonic beers were becoming well known and their popularity continued through to the nineteenth century. Some were used to replenish the body's vitamins which had become run down with the shortage of fresh fruit and vegetable during the winter. They were known as spring drinks. Culpeper in his herbal of 1652 gives details of a tonic beer made from black alder.

The prevention of scurvy was one of the main health concerns and the best known herbs for preventing this at the time was scurvy grass, which was the reason for its name. The household accounts of the Russell family, in 1653, must be one of the earliest references to scurvy grass ale. As an item in the accounts, scurvy grass was shown to be bought for putting into the children's beer. The drink had become a little more sophisticated by the eighteenth century as the following recipe shows:

Alexandrian Sena, freed from stalks	4 oz
Rhubarb sliced	1 oz
Winter's Cinnamon	3 oz
Polypody of the oak	1½ oz
Bay and Juniper berries, of each	1 oz
Anise and fennel seeds, bruised, of each	1 oz
Liquorice and Horse Radish, sliced	1½ oz
Seville Oranges cut in pieces	6

Put all the ingredients into a bag, with a stone in it to make it sink, into three gallons of ale; take a pint and a half of the juice of garden scurvy grass, set it over the fire and clarify it, let it stand till cold, and then put it into the ale.

> After it has worked together for a day and a night, stop up the vessel close, and when it has stood six days, drink a pint warm tasting.

Other recipes also used cress, brooklime and sage.

By the nineteenth century scurvy grass ale was being produced commercially in larger quantities. It was then more practical to add the herbal ingredients during the brewing process. The ingredients for a brew of 5 quarters of malt included:

Hops	25 lb
Molasses	10 lb
Scurvy grass, to the copper	5 bushels
Alexandrian Senna, to the copper	2 lb
Horse Radish root, to the F.V.	1 lb

The beer produced from the recipe was not very strong as all the wort recovered was used for making the ale, whereas the strong ales of the time only utilised the strong first batch of wort produced; the weaker wort from the second or third batch was used for making small beer. In the light of what we know today, it will be appreciated that the beer would have had very little value in preventing scurvy, as the Scurvy grass was boiled in the copper which would have destroyed the active ingredient, vitamin C. Other methods of production in which the herb was added after fermentation would of course have been effective.

Another recipe of the time used no hops but instead included the bitter herbs Roman Wormwood, Gentian and Orange peel.

An interesting concoction though not strictly an ale but a use of ale in an early type of anaesthetic is found in the *Natural History of the Herring*, 1753 by Dr Silas Dodd. He gives the following instructions:

> Take the oil pressed out of fresh herrings, a pint, a boar's gall, juices of henbane, hemlock, arsel, lettuce and wild catmint, each six ounces, mix, boil well, and put into a glass vessel , stoppered. Take three spoonfuls and put into a quart of warm ale, and let the person to undergo any operation drink of this by an ounce at a time, till he falls asleep, which sleep he will continue the space of three or four hours, and all that time he will be unsensible to anything done to him.

At the end of the eighteenth century spruce beers were popular medicinal ales. They were prepared from the the extract of young spring shoots and toppings of the Red, Yellow and Black Fir trees growing in the northern parts of Europe, Asia and America. The autumn cones were also used. The Fir trees of more southern regions could not be used as the flavour was not suitable, verging on the flavour of common turpentine. The extract only provided flavour, the fermentable matter coming from molasses. Typical proportions for 20 gallons of spruce beer were:

10 lb Molasses
1 lb Spruce essence

Table beer was also used to produce spruce beer and sugar could be used to produce a white spruce beer. Its medicinal properties were described as follows:

> All preparations from spruce are warm, attenuating and more or less stimulating and antiscorbutic; promote urine, sweeten and purify the blood and animal juices, cleansing and heating internal ulcerations, particularly those of the urinary passages; at the same time, like other bitter hop substances, strengthen the tone of the vessels, and are particularly good in gravelly and other calculous complaints, dissolving mucous, and promoting the expulsion of sand; in gleets and seminal weaknesses are good diuretics, and accelerate the cure.

It was also believed that it could be used to prevent yellow fever.

Another medicinal beer of the time was cherry beer, which was made by adding a few pounds of molasses to a barrel of table beer with a pound or less of logwood and two ounces of almond cake. There were no cherries used; the logwood gave the red colour and the almonds the flavour. It was less popular than spruce beer which was a laxative because it had the reverse effect.

A medicinal ale known as Panala was one of those produced in the nineteenth century. Its ingredients were not known but it was described as thus:

> This ale neither offends the eye with the loathed object of a muddie substance, nor the smell with any ill vapour or favour, nor the tast nor stomacke with disgust or ingrate relish, but 'tis a pure, cleere, delicate, and singular extract impregnated with the sincere spirits and vertuosities of excellent ingredients, of a moderate temperature, indifferently accommodated to every age, sex, and constitution, and so familiar and pleasing to nature.

Another noted medicinal beer of the time was Dr Butler's Ale which maintained its popularity for many years. Bickerdyke gives the following details:

> The following receipt for this ale is given in the *Book of Notable Things*: Take senna and polypedium each four ounces, sarseperilla two ounces, agrimony and maidenhair of each a handful, scurvy grass a quarter of a peck, bruise them grossly in a stone mortar, put them into a thin canvas bag and hang the bag in nine or ten gallons of ale; when it is well worked and when it is three or four days old, it is ripe enough to be drawn off and bottled . . .

> It is an excellent stomach drink, it helps digestion, and dissolves congealed phlegm upon the lungs, and is therefore good gainst colds, coughs, ptisical and consumptive distempers; and being drunk in the evening, it moderately fortifies nature, causeth good rest and hugely corroborates the brain and memory.

BOTANIC BEERS

Herb and spice beers continued to be made throughout the nineteenth century and by the end of the century their commercial production was in the hands of a specialist type of brewer known as the botanic brewer. This was perhaps because there had been a steady decline in the amount of domestic brewing which had previously been the main supply of such beers. The shift of the population into towns and cities meant that many people were unable to gather wild herbs or cultivate them even if they did brew their own beer. There was also an increasing demand for such beers because of the temperance movement. They were not opposed to such beers, because of their medicinal or tonic value and because by nature they were never very strong. The only reason that they were fermented at all was supposed to be because the carbon dioxide produced gave them a natural effervescence. It was a time before the technology of artificial carbonation had been developed. The fermentation time was only twenty-four hours, which was sufficient to give the amount of effervescence required, but in theory very little alcohol. The Customs and Inland Revenue Act of 1885 decreed that such beers, if they were under 2% proof spirit, were not liable to duty and could be sold without a licence. This rather erroneously lead to them being called non-alcoholic beers. The fact that some people could mildly indulge in the consumption of alcohol and still maintain a façade of temperate respectability must have helped their sales.

It was perhaps the difficulties of maintaining the alcohol content below 2% which lead to their demise. H. M. Customs and Excise were well aware that they were a back door to producing duty free beer so kept a very firm control on their production. Several prosecutions were made; James Stanger in Sheffield was prosecuted for producing a Ginger beer with a content of 10.8% proof spirit. Guinness stout at the time had a content of 11.0% proof spirit. These and other production problems lead to the eventual decline in botanic beer sales and after World War I manufacturers concentrated on what are today known as soft drinks using artificial carbonation.

The more common botanic beers included: Ginger beer, Burdock ale, Dandelion stout, Dandelion and Burdock ale, Camomile beer, Horehound beer, Lemon beer, Spruce beer, Sarsaparilla beer and various concoctions which were merely called herb beer. The way in which they were made was quite simple compared with malt beer brewing. The herbs or spices were boiled for fifteen or twenty minutes in order to extract the flavouring substances. The mixture was strained and poured on to sugar and saccharine, the amounts being carefully calculated to give the right degree of fermentation and sweetness. Cream of tartar or tartaric acid was usually added with any necessary colouring. When cool, yeast was added and after a short fermentation it was bottled.

The only botanic beer which has survived is Ginger beer and in 1932 it was the only soft drink which was allowed to include the word beer in its title. Other botanic beers were made into soft drinks such as Dandelion and Burdock.

A recipe for home made Ginger beer of 1853 is:

3 lb White sugar
1 oz Bruised ginger (or $\frac{2}{3}$ oz ground)
1 oz Cream of tartar
Juice of 4 lemons

Soak the ingredients in 4 galls of boiling water for two hours, strain, cool to 60°F. Add yeast or ginger beer plant and ferment for four days or until the fermentation is finished, then bottle. The sedimented yeast is divided for re-use or to pass on to others as a ginger beer plant.

Root beer was a popular beer in the nineteenth century, particularly in the U.S.A. The flavouring ingredients in one recipe included: bruised sassafras bark, bruised wintergreen bark and bruised sarsaparilla root.

NATIVE BEERS

It is of interest to note that all the adjunct cereals are used, or have at some time been used, as the main material to produce native beer or beer which is peculiar to a particular country, region or even tribe. There are many regions of the world where barley cannot be grown and the traditional European style of beer could not be produced even if there were a desire to do so. Through colonisation and western influence European style beers have been introduced into some countries where barley is not a native crop and displaced the beers made with indigenous materials. One country where this has happened is Nigeria and such has been the success of their European style brewing industry, that the importation of European brewing materials has put their economy under a heavy strain. Their government has been forced to restrict their importation in the future which will have a profound influence on

TRADITIONAL AFRICAN
BREWING MILLETS
SORGHUM IN CENTRE

the methods of brewing in that country, doubtless forcing their brewers to look again at their native beers.

In Africa most native beers are made with millet. The type of millet which has the largest grains and the greatest economic importance is sorghum. It is known by a number of names: in the Sudan and Egypt it is known as durra, in West Africa as guinea- corn, in East Africa as mtama and in South Africa as Kaffir corn. The many different types of beer produced are too numerous to mention in this book but it will serve to take examples from each region.

In the area known as upper Egypt a native beer made from durra was Bilbil, or Omm Bilbil, which derived the name from the native word for 'the mother of the nightingale'. This was because it made drunken people break out in song. In its production the grain was brought to germination between large leaves, dried in the sun and then finely crushed. It was then put into large earthenware pots, boiled for 6 to 8 hours and the mash allowed to cool. Yeast was added and fermentation proceeded. The mash was then filtered, boiled a second time and fermented again to produce a beer resembling sour champagne.

In Nigeria one of the traditional native beers made with Guinea-corn is called Otika.

In Southwest Africa a native beer of the Ovambos called omalofo is made from Kaffir corn. The traditional method of production is first to soak the corn in water and then to dig it into the ground in order to germinate it. It is then dried, made into a coarse flour, boiled with water, sieved through palm leaves and pharmaceum grass and the clear fluid run into clabashes. In these it is stored and allowed to undergo a long fermentation before drinking. Kaffir corn along with millet and maize was also one of the ingredients used to make the native Kaffir beer of the Matabele, in what is now Zimbabwe. In this country Kaffir beer is now produced from sorghum on a commercial scale comparable with the industry in Europe.

Other native African beers were also brewed with millet and maize, such as Dolo, a beer of what was French Sudan. One of the African tribes noted for their brewing were the Wahehe of East Africa who produced a beer called Pombe from millet. They had special brewing and drinking houses and indulged themselves in drinking on a big scale. Pombe was the name used for native Kaffir beer over all southern Africa and this too is now produced on a large commercial scale from sorghum. It is now preferable to call it Bantu beer as Kaffir is regarded a a somewhat derogatory word. Calling the beer Kaffir beer would be as tactless as insisting on calling German beer Kraut beer.

A POMBE ORGY

In Central America and South America, cassava (manioc or yuka) root is a valuable source of starch and beers of various kinds are made from it. The basic method of production is to chew the root, or a bread made from the root and to spit it out into a container of water; the enzyme ptylin in the saliva then converts the starch to a fermentable sugar from which the beer is produced by spontaneous fermentation. This requires the labour of several people in order to produce enough material for a brew of any size and the task befell children and women in the communities in which it was made. Examples of such beers are Perrione produced by the Tayos of Guiana, Piworree produced in Barbados, Ouyou produced on other Caribbean Islands and Masato produced in Peru.

In the Far East rice is used for brewing and China and Japan are particularly noted for their rice based drinks. There seems to be a general preference to call these wines rather than beer because of their delicate flavour, but technically they are beer as they are derived from a starch source rather than a sugar source. Shaohsing Chia Fan Chiew is the most famous Chinese rice wine and is brewed from rice and millet. It is known to have been brewed for the last 2,000 years. Wu Liang Ye is another noted rice wine which is literally translated as 'five grain liquid'. It is made from rice, glutinous rice, sorghum , corn and wheat. Kiu is a very ancient Chinese beer made with millet and it is mentioned in two ancient books that are between three and four thousand years old.

All the aforementioned beers are more or less regarded as wholesome drinks but one particular native beer made from Kava root has produced some very harmful effects on its drinkers. It was again made by masticating the peppery-flavoured root. It was encountered by Captain Cook on his voyages in the Pacific in the eighteenth century. Members of the crew who visited Tahiti on more than one visit, noticed that some of the men who had previously been in good health, had by the time of the second visit developed dry rough scaly skins and were almost reduced to skeletons. This was attributed to the drinking of the beverage made with kava. A visitor to Hawaii in 1821 where the beverage was also drunk made the following observation:

> When a man, says he, first commences taking it, he begins to break out in scales about the head, and it makes the eyes very sore and red; then the neck and breasts, working downwards, till it approaches the feet, when the dose is reduced. At this time the body is covered all over with a white scurf, or scale, resembling the very dry scurvy. These scales drop off in the order of their formation, from the head, face, neck, and body, and finally leave a beautiful smooth clear skin, and the frame clear of all disease. The process is also held to be a certain cure for venereal infection.

CHAPTER 9

LEGISLATION AND CONTROL

FOOD LEGISLATION IN BRITAIN

In the United Kingdom, as in many other countries, the health of the food consumer is protected from the effects of additives, adulterants and contaminants by various pieces of legislation. Legislation also, to some extent, protects the consumer so that when he buys food he gets fair value for his money and is not financially exploited by the use of adulterants. The situation in Britain is that detailed control of additives and contaminants is carried out through various regulations made under the Food Act 1984, Part 1, Section 4 (1). National legislation is supported by EEC regulations which have been introduced mainly in order to harmonise the legislation of Member States. This assists the movement of goods within the EEC so that trade is not hindered. The process under which legislation is introduced within the EEC, as would be expected, is extremely complex but ultimately a Directive, as the piece oflegislation is known, is adopted by the Council. It is a very flexible instrument merely stating objectives and it is left to the Member States to decide how they achieve them.

DEVELOPMENT OF LEGISLATON

Modern legislation in Britain started in Victorian times with the Adulteration of Food and Drink Act 1860 and the Adulteration of Food, Drink and Drugs Act of 1872 which mainly required local authorities to appoint public analysts and empowered inspectors to take samples for analysis when required. In 1875 the Sale of Food and Drugs Act was passed in which were embodied the words: 'No person shall sell to the prejudice of the purchaser any article of food or anything which is not of the nature, substance or quality demanded by such purchaser.'

In 1925 regulations were brought in which prohibited the use of preservatives in certain foods and in 1938 the Food and Drug Act empowered Ministers to prohibit the use of several different groups of additives. However, it was only in the Food and Drug Act of 1955 that the modern legislation as we now know it was introduced. The latest consolidation of this Act took place with the Food Act of 1984.

The Regulation made under these Acts which affect beer include the following:

The Preservatives in Food Regulations 1979 (S.I. No. 752) with amendments 1980 (S.I. No. 931) and 1982 (S.I. No. 15).

The Miscellaneous Additives in Food Regulations 1980 (S.I. No. 1834) with amendment 1982 (S.I. No. 14).

The Antioxidants in Food Regulations 1978 (S.I. No. 105) with amendment 1980 (S.I. No. 1831).

The Emulsifiers in Food Regulations 1980 (S.I. No. 1833) with amendments 1982 (S.I. No. 16) and 1983 (S.I. No. 1810).

The Colouring Matter in Foods Regulations 1973 (S.I. No. 1340) with amendments 1974 (S.I. No. 1119), 1975 (S.I. No. 1488), 1976 (S.I. No. 2086) and 1978 (S.I. No. 1787).

Some additives and processing aids used in brewing , such as isinglass, enzymes and substances used in liquor treatment are not covered by these regulation. However, the consumer is protected by the general provisions of the Food Act which prohibits the use of ingredients in the production of food which would render it injurious to health.

The way that Regulations come into force is that they are drafted by the Ministry of Agriculture, Food and Fisheries and then put before Parliament for approval, before being made law. They do not have to be debated but can be challenged by Members of Parliament. This rarely happens.

APPROVAL OF ADDITIVES

There is an approval procedure for the acceptance of additives for use in food. Some of the main details are outlined:

The MAFF is advised by two main expert committees: The Food Advisory Committee (FAC) and the Committee on Toxicity of Chemicals in Food Consumer Products and the Environment (COT) which reports to the FAC. The FAC is an amalgamation of the Food Standards Committee (FSC) and the Food Additives and Contaminants Committee (FACC) which came together in 1984. The members of these committees are chosen by the MAFF and there has been some comment that the members who are chosen are people who are unlikely to want to alter the situation from the way that it is seen by the Government. The representatives of industry far outweigh those of consumers. However, the committee is basically one of experts rather than one of representatives of interested groups. There are perhaps more experts in industry and scientific establishments than there are in consumer groups.

Very often, when the use of additives is reviewed they are considered in groups, as was the case when beer additives were reviewed in 1978. When such a review is about to commence interested parties are notified through a press release. Information is submitted from various sources on the additives under consideration. This includes data on their rates of use, specifications and details of their safety in use. The two main points which are considered are the necessity of the additive and its record of safety in use. The advice on the later is taken mainly from the COT but advice is also taken from the Committee on Carcinogenicity (COC) and the Committee on Mutagenicity (COM) which are committees of the DHSS. Their activities are covered by the Official Secrets Act but their reports usually appear as an appendix in FAC reports. The FAC is not affected by the Official Secrets Act.

On publication of the FAC report another press release is issued inviting comments on the report and these are duly considered before proposals for regulations are made. A further opportunity is then given for comments to be made and considered. The comments on proposals usually come from industry rather than consumer groups although they have an equal opportunity to make them.

Applications to use an additive may be made at any time as well as at a general review and the same consideration is given to its use.

The EEC has its own Scientific Committee for Food (SCF), on which the U.K. is represented. It makes recommendations to the EEC Commission so that EEC legislation can be formulated. Not all food additives are the subjects of Directives but by the 1990s it is expected that evidence collected by the SCF on the acceptability and need of all food additives will enable legislation to be made.

The EEC has given a number, the 'E Number', to most approved additives. These are used particularly in food labelling. Those substances without an 'E Number' but permitted in the U.K. are just given a number under the Food Labelling Regulations.

BODIES CONCERNED WITH FOOD SAFETY

There are other Government Committees who advise on the substances which may affect the safety of food. One of these is the Steering Group on Food Surveillance. It was formed in 1971, made its first report in 1978 and by 1988 had published twenty-one Food Surveillance Papers, most of them covering particular food contaminants. Some of the substances looked at are found in beer.

There are various international bodies which provide advice on food additives. One of the most important is the Joint Expert Committee on Food Additives (JECFA) of the United Nations Food and Agricultural Organisation/World Health Organisation. This group of international experts evaluate the latest information on food additives and make recommendations for their use. Their reports are of assistance to government committees when considering the use of additives in the U.K.

It was through the JECFA that Acceptable Daily Intake (ADI) levels for additives were first introduced. This is the maximum number of milligrammes of an additive which can be consumed per kilogramme of the consumers' body weight. This is obtained by measuring the largest dose which gives no adverse effects on animals. Taking into account that humans may be more sensitive than laboratory animals and that there is a natural variation between humans, a hundred-fold safety factor is allowed. These ADI figures are not, however, used in U.K. Food Regulations although they are used in other European countries. They are very difficult figures to use in order to set the maximum amount of an additive which should be allowed in a particular food, as there is no control on how much of the food is consumed and how much of the additive is taken in with other foods.

CONTROL FROM WITHIN THE INDUSTRY

The brewing industry takes a very active interest in the additives and contaminants of beer and has taken numerous measures on its own initiative in order to make beer safer to drink. The trade association of the brewing industry is the Brewers' Society and most of the larger old-established breweries belong to it. It is funded by member companies who pay according to the barrelage that they produce. It represents the industry as a whole, when necessary, in negotiations with the Government, the EEC and anyone else who needs to communicate with the industry. Policy is decided by the Council of the Society, who receive reports from various committees. If reports received are approved they become Society policy and approved reports are circulated as confidential bulletins to members.

Food safety in the Brewers' Society comes under the control of the Technical Committee. A sub committee called the Scientific Sub Committee has the main function of monitoring the safety in use of all materials, additives and processing aids used in brewing and also materials which come into contact with beer such as those used in packaging and in the dispensing of beer. Much work on food safety is carried out at the Brewing Research Foundation which is largely funded by the Brewers' Society. They have representation on the Scientific Sub Committee.

The Brewing Research Foundation operates a Food Safety Watch. Information is gathered from a large number of international sources including: literature on brewing and related disciplines, computer data bases, Government publications, legislation, trade literature, trade journals and newspapers. Details of the information are recorded in a computerised information retrieval system for the use of the industry.

Through the Technical Committee the Society disseminates important information to members and makes responses to the information when necessary. Examples of action taken by the Brewers' Society in order to make beer safer to drink are given below:

The Agricultural Chemicals Working Party which reports to the Scientific Sub Committee arranges for the testing and assessment of chemicals such as pesticides used during cultivation and storage of brewing materials. A list of acceptable chemicals has been prepared and is regularly updated. In 1981 the Brewers' Society was instrument in limiting the amount of bisdithiocarbamate residues in hops. The main degradation products of these, which are produced in beer, are thioureas. It was found in 1978 that they may be carcinogenic, mutagenic and teratogenic

Extensive efforts have been made to reduce the amount of nitrosamines in malt and beer, including contributions to research and extensive monitoring.

In 1978 EEC legislation caused the permitted lead levels in beer to be lowered and on the recommendations of the Brewers' Society the industry spent several million pounds modifying beer engines so that they were made of materials which contained no lead.

The Brewers' Society recommended to members that formaldehyde should not be used for brewing in advance of public concern over its use.

Recommendations by the Brewers' Society are not mandatory for members but the value of the recommendations are generally recognised and the indications are that they are followed by everyone. Virtually no brewer would be prepared to risk his reputation by being found to be out of line with the rest of the industry in the field of product safety.

CONSUMER PRESSURE

The main consumer group for beer drinkers is the Campaign for Real Ale (CAMRA). As their title suggests, they are fighting for the preservation of real ale, which is defined as beer that has not been artificially carbonated to give it high levels of dissolved carbon dioxide. The addition of this extra carbon dioxide is seen as adulteration, as its sharp taste masks other flavours in the beer. It can be used to cover up shortcomings in a beer as it makes the flavour more interesting, though not necessarily more desirable. Some artificially carbonated beers taste very insipid when their carbonation levels are reduced to those of naturally conditioned beers. It is also very uncomfortable for the drinker when carbon dioxide gas is released in the gut after drinking.

CAMRA has widened its interests to include other issues which involve the beer drinker. They are generally concerned about the use of additives as they believe that they can present hidden dangers to health and that short-term safety trials carried out on animals are not necessarily relevant to ourselves. Interaction between additives and the effects of allergies on a minority, they say, are also rarely considered. The campaign would like the introduction of product labelling of ingredients for beer and regard avoiding the use of unnecessary ingredients as essential.

CAMRA are against the nationally distributed beers, as they prefer a greater number of locally produced beers which give the consumer more variety. Additives are seen as aids to the national brewers enabling them to produce standard beers able to withstand extensive distribution over longer periods. The use of additives is therefore considered in this case to be in the interests of the big brewers rather than the consumer.

CAMRA feel that they have increased the awareness of beer drinkers to the use of additives and have informed them of possible dangers or malpractices. Their research has been limited because of the high cost of chemical analyses but a survey carried out in 1987 revealed high levels of some metal ions, which were above the EEC limits for drinking water. The small sample of a particular brand of mild ale showed evidence of the practice of adulteration by the addition of water. There was no other evidence of adulteration, contamination or misuse of additives presented.

PURE BEER LEGISLATION IN BRITAIN

It is now generally accepted that in Britain beer is brewed with materials other than malted barley and hops, but this has not always been the case. In 1802, Britain introduced legislation prohibiting the use materials other than malt and hops which backed up similar legislation passed in the eighteenth century. The main reason for the legislation was not to protect the customer, but to protect the Revenue as there was a tax on malt and hops. During the nineteenth century there were a number of legislative changes which gradually allowed other materials to be used and these are detailed in Appendix 2. Initially, these changes were minor and allowed the use of colouring materials and the temporary use of sugar. The first major change was to allow the use of sugar on a permanent basis in 1847. The next change was to abolish hop duty and allow the use of hop substitutes, but this only slightly affected the use of hops because of their superiority as a bittering ingredient. The biggest change occurred in 1880 when the Free Mash Tun Act was passed repealing malt duty and allowing the use of malt substitutes. In Mr Gladstone's own words: 'The brewer will brew from what he pleases and have a perfect choice both of his material and his methods.'

THE PURE BEER BILLS

The Free Mash Tun Act resulted in two main areas of concern. Firstly, there was concern that the brewers would use deleterious materials and secondly, that the trade in barley would be adversely affected through the use of imports such as sugar, rice and maize. In order to protect these interests a number of Pure Beer Bills were brought before Parliament in an attempt to introduce restrictive legislation.

There was no real reason to believe that brewers would see the Free Mash Tun Act as a chance to adulterate their beer, as the public had the protection of the 1875 Sale of Food and Drugs Act. Legislation was also passed in 1885 which specifically prohibited deleterious substances being added to beer by brewers. The production of beer unlike any other food was also carried out under the eye of the Excise, who were always on the look out for the use of deleterious substances. In truth, the main reason why these Bills were brought to Parliament was through pressure from the arable farmers who saw a large slice of their sales likely to disappear. They perhaps saw the problems which had been encountered by the hop growers when hop duty was repealed and they faced competition from substitutes. There was some pretence that the Bills were for the protection of the consumer as their titles seem to imply, but a writer of the time insisted that the title of one of the bills was a misnomer and that it would be better renamed the 'Barley Growers' Protection Bill'.

The first Bill was introduced by Colonel Barne who represented a constituency in the rich barley growing area of East Suffolk. He wanted people who sold beer, if their beer was made with materials other than malted barley or hops, to conspicuously post a notice stating what other materials were being used. The Bill failed, but in the same Parliamentary Session Mr Chaplin who looked after farmers' interests in the House made the proposition:

> That this house is of the opinion that the repeal of the malt tax, by encouraging the use of rice, maize and other articles as substitutes for malt in brewing, has proved to be injurious, instead of beneficial, to the farmer; that the manufacture of beer of inferior quality from such articles is contrary to the public welfare, as well as detrimental to the interests of agriculture, and that no financial scheme will be satisfactory which does not amend the law with regard to the use of ingredients other than sugar, malt and hops, in the brewing of beer for sale.

He had been for many years an advocate for the repeal of the malt tax, but like the hop growers several years before who petitioned for the abolition of their duty, he changed his mind when the effects of such legislation were realised. The proposition was rejected by a large majority. Parliament also effectively dismissed another proposition brought by Mr Heneage which was similarly worded.

In 1885, 1886 and 1887 several attempts to get a Pure Beer Bill through Parliament were made. The main attempt seemed to be by Mr Quilter whose bill received a second reading in 1886 but had to be reintroduced in 1887. The attempts were heavily opposed by an organisation called 'The Free Mash Tun Association', led by Mr Stopes a consultant brewers' engineer and maltster, and backed by many brewers. Just before Mr Quilter's Bill was due to have its second reading in 1887 a group of supporters which included barley growers, hop growers and several Members of Parliament met with Mr Goschen, the Chancellor of the Exchequer, in order to put their case. He told them that the weakness of their case was that they had no evidence that the use of materials other than malt or hops are injurious to health and if they wished to prove that beer brewed from other materials is bad beer, they would have to find evidence. A rather scathing interpretation of the proceedings made by an unsympathetic observer was, 'that the Chancellor metaphorically kicked their feeble arguments round the room and then kicked the deputation out of it'. Not surprisingly Mr Quilter had no success with his Bill and it was not even brought before the House for discussion on the day it was due for its second reading.

Despite the failure of his Bill Mr Quilter continued to revive it every year until 1896, when again it came up for a second reading. At this stage he was persuaded by the Chancellor of the Exchequer to drop the Bill on the promise that a Parliamentary committee would examine the practicability and desirability of legislation along the lines suggested.

In order to support the cause of the Pure Beer Bill, petitions had been presented regularly between 1884 and 1896. Between 1884 and 1890 there were ninety-one presented with a total of 19,841 signatures; no more were collected until 1896 when there was a big push and forty-seven petitions were collected with a total of 28,861 signatures. There were also fifty three informal petitions which were rejected by the Select Committee on Public Petitions. It was admitted that the petitions were collected mostly in East Anglia and some in the West Country, but they were said to have been signed by people in all walks of life, not just by farmers. These petitions which had been signed by about 0.5% of the population, coupled with the fact that Mr Quilter was now Sir Cuthbert Quilter, doubtless influenced the Government to set up the promised select committee.

INVESTIGATIONS OF THE SELECT COMMITTEE

The Committee was to report on the question:

> Whether legislation is required to prevent the use of deleterious substances in the manufacture of beer; and whether the materials of which beer may be composed can be defined by law without undue interference with the liberty of brewers to use any wholesome materials in brewing.

Sir Cuthbert Quilter's concern over the use of adjuncts was for the detrimental effect it was having on agriculture and also for the detrimental effect it had on ' . . . the health of the public and the fairness to the individual, that he shall have what he believes himself to be buying'. It appeared during the giving of evidence to the Committee that Sir Cuthbert was putting more emphasis on the concern for the consumer than the need to protect the farmer. This may have been that he saw it as an approach more likely to succeed after hearing the evidence that had been given before his own. He had the advantage of being one of the last to give evidence, which was taken over twenty-six days from a variety of witnesses who included members of the Inland Revenue, public analysts, brewers, farmers, hop growers, manufacturers and various consultants.

In its report the Committee found that they could not recommend any interference with the agricultural industry, as they felt it unlikely that the use of adjuncts affected the demand for British barley. They considered that prohibiting the use of sugar, which was used to decrease the nitrogen content of beers and the risk of haze formation, instead would increase the demand for foreign barleys which had lower nitrogen contents. The Committee also decided that there was no threat to the health of the public from materials used by brewers. None of those reported as being used were considered deleterious or likely to adversely affect the nutritive value of beer.

Sir Cuthbert Quilter was not only asking for the prohibition of additives, but that the public should be made aware of the ingredients of any beer made with materials other than malt and hops and that they should be indicated on labels and on notices at the point of sale. He asked for a definition of 'pure' beer so that the public could obtain a product brewed from what they expected it to be brewed with. If they wanted a product brewed with other materials, which could be described by another name (such as 'beerine' liquor), then they could have the choice. Although the committee was said to appreciate these views, it felt that with the state of scientific knowledge being what it was, it was impossible to determine what type and how much of an additive had been added to beer. It considered that it was undesirable to pass a law which could not be upheld because it would be impossible to prove when it was being broken.

If these were the true grounds for rejecting the labelling of beer perhaps if a Bill were introduced today it would have a high chance of success with the analytical techniques now available.

THE REINHEITSGEBOT

The Reinheitsgebot is the German beer purity law which prohibits the use of any materials in brewing other than malted barley, hops, yeast and water. It was introduced in Bavaria, in the sixteenth century, but legislation concerning beer had been in existence in Germany for several hundred years before this time.

THE HISTORY OF LEGISLATION

In Hamburg, in 1276, brewing was made a specific civic trade which meant that it was subject to ordinances and decrees (Burspraken) which were publicly proclaimed each year. From the fourteenth century nearly every Burspraken dealt with the brewing industry, illustrating the importance which was attached to its control. The Town Fathers were aware of the importance of the industry, not only because of the revenue which it brought in, but also because of the trade which it attracted to the town. The concluding words of many public speeches were often: 'May God protect the brewing industry of this town'. The words were also an ancient church prayer.

Other German towns such as Lubeck, Rostock, Hanover, Cologne, Erfurt, Nuremberg and Munich also exercised some degree of municipal control by the promulgation of various regulations.

The legislation was all concerned with making the brewing industry of the town competitive in order that it might prosper. It covered various aspects, such as who could brew, when they could brew, what materials could be used, what amounts of materials could be used and even the method of brewing. In short, any aspect which affected the quality of the beer was covered in some way. Fairly early on in the history of German brewing it was established which materials produced the best beer. After a period when gruit was the most popular flavouring material, the superiority of the hop was realised and by the end of the fifteenth century gruit was specifically prohibited. Barley was generally found to be the most practical cereal with which to brew, although several towns produced a wheat beer in which wheat was a portion of the grist. However, as early as 1351 Erfurt for example prohibited the use of oats and wheat in order to preserve the character and quality of their beer.

The part of Germany where this type of legislation was brought in with the greatest force was Bavaria where all ingredients with the exception of malted barley , hops and water were prohibited. It was first introduced in 1487 by Duke Albrecht IV in Munich itself and in 1493 by Duke George of Bavaria-Landshut in Landshut which is about 75 km to the north of Munich. Eventually all 'Old' Bavaria was brought under the legislation by Duke William IV in 1516 at the Diet of Ingoldstadt. Bavaria has always adhered to these laws, which achieved their original aim, of giving Bavaria an excellent reputation for its beers. Such was the success of the law that the whole of Germany was persuaded to follow Bavaria's example, though at a much later date. Badenia accepted the law in 1896, Wurttemberg in 1900 and eventually the North German Beer Tax Convention on the request of the whole of the German brewing industry in 1906.

PROVISIONS OF THE REINHEITSGEBOT

To say that the law only allows the use of malted barley, hops, water and yeast is a simplification of the law as it stands. They are the only materials allowed in the brewing of bottom fermented beers but wheat can also be used in top fermentation beers. These are becoming increasingly popular, especially with younger people as a refreshing summer drink. In 1984 for example, in Bavaria they represented 17% of the market, with rising sales against a fall in beer sales of 2% as a whole for that year. A category of beers known as Special Beer can be brewed with beet, cane or invert sugar and colouring matter made from sugar. These are mostly produced in the north of Germany where they only represent about 1.5% of total production in 1985. Beers brewed for export do not have to comply with the Reinheitsgebot although they do have to comply with other food regulations. Bavaria, Badenia and Wurttemberg, however, do apply the Reinheitsgebot to their export beers.

Some other substances may be used in brewing if it is considered that they do not pass into the final product. These include substances for removing haze forming material such as silica gel, bentonite and PVPP. Chemicals may be used to treat the brewing liquor as in the U.K., with certain restrictions. Gypsum and calcium chloride can be added provided that they do not exceed the amounts found in natural water supplies. Malt has to be made without the use of additives, although barley may be washed during the early stages of steeping with alkaline cleaning agents, but these have to be washed off in the ensuing processing. Smoked malt is used in the Bemberg area and is made by burning beech logs in the malt kiln. It has caused some problems with controlling the amount of nitrosamines in the beer.

The amount of sugar used in producing the Special beers of Northern Germany is considerable. Today the grist is 100% malt brewed to a gravity of 8°P and the sugar is added after a short fermentation to give a gravity of 12°P. In Bavaria they cannot be sold as beer but are designated 'malt beverages with sugar'. In N.W. Germany sugar and saccharine are used to produce beers known as 'Sweet' or 'Caramel' beer. They are traditional beers which are not allowed to be sold outside the area.

Hop extracts are allowed provided that they contain the whole extract of the hop and nothing else. Solvents such as methylene chloride used for extraction should not remain in concentrations above 0.1%. Isomerised extracts which are added after boiling are not permitted.

Washing of yeast in order to remove bacteria as practised in the U.K. is allowed, but only sulphuric acid can be used and has to be completely washed out afterwards.

INFRINGEMENTS

The Germans have always taken great pride in producing their beers according to the Reinheitsgebot and most rigidly enforce it. On the infrequent occasions when German brewers have been caught infringing the law the public condemnation and subsequent disgrace offers a very bleak future for the brewery and the careers of the brewers involved. The most recent case of infringement, in 1985, is known as the Kieninger Scandal. A number of German breweries, particularly in Bavaria, were alleged to have added monobromo and monochloro acetic acids to cure beers which had been spoilt during processing. These poisonous substances were said to have been recommended by a well known brewing consultant Professor Kieninger, who also sold the chemicals to the brewers. Though the brewers were infringing the Reinheitsgebot they were tried on infringement of the food safety laws, which operate for all foods in Germany. Professor Kieninger who had been arrested but allowed bail, on returning from a business trip to the Far East had his bail bond revoked on the grounds of obscuring evidence. Sadly, the Professor preferred death to dishonour and one week later killed himself in a jail near Munich. The brewery companies and their head brewers were fined and the beer was put down the drain. The first brewery to be prosecuted was fined £12,500 and the head brewer £555, but the real punishment was their loss in reputation and its effect on beer sales for the brewery.

It is a requirement of German law that beer imported to Germany should comply with the Reinheitsgebot. This was perhaps a little difficult to ensure in the past but with modern analytical methods it is easier. Immunoelectrophoresis and gel diffusion techniques were developed to detect the presence of enzymes and adjuncts in beer. In a survey of ninety imported beers in 1978, 43% were found to contain papain and the cereal adjuncts wheat, corn and rice were also detected. The beers were imported from Czechoslovakia, France, Denmark, Netherlands, the U.K., the U.S.A. and East Germany.

THE VIEW OUTSIDE GERMANY

As some parts of Germany are allowed to use additives in their export beers many countries have felt that the Germans are being hypocritical. It is a particular point which has been irritating the French for a long time. As far back as 1886 they were getting very concerned about the amount of salicylate of soda and salicylic acid in imported Bavarian beer. It was said that 'there was not a doctor in the world who would prescribe half as much to be taken every day for an indefinite period as there was in the Bavarian beer'. The amount used was estimated to be 300 mg/l. It was discovered that the Germans were not being subjected to these chemicals in their own beer, as their government prohibited their use at home, but allowed their use in beer sold to the French. However, there was little that could be done about the matter until

Germany became part of the EEC. Under EEC regulations any product legally made and sold in one EEC country can be traded with any other EEC country. A formal complaint was lodged with the EEC Commission in 1982 by the Brasseries du Pecheur of Alsace, whose beers had been refused for import into Germany. They were particularly aggrieved that the German laboratories showed that their beers contained additives despite their claims that none had been added. The French also saw Germany as a particularly attractive market on their doorstep which consumed 2.84 billion gallons of beer. Only 23.8 million gallons were imported whilst 132.1 million gallons were exported in 1986. The complaint was also lodged against Greece who was exercising similar restrictions.

The Greeks gave way but the Germans were determined to make a stand. Their main argument was that the law was necessary on health grounds, as the Germans drank so much beer as part of their diet. They claimed that they would be consuming extra amounts of additives from which they would suffer harmful effects. The following points were raised against the German argument. The Reinheitsgebot is part of their beer tax law, not part of their food safety law, which like other countries exists to protect them against additives in all foods. The Germans' concern for food safety with beer only extends to their own people as their export beers can contain additives prohibited in Germany. They are allowed to use some selected processing aids in their own beer which are prohibited in other countries. One of these is PVPP which is prohibited in France as residues are believed to be potentially carcinogenic. Traces have been found in imported German beers by the French. All EEC countries have strict food safety laws which prevent them exporting beers which contain dangerous additives and need no other safeguards. The German claim that they are at special risk because of their high per capita intake of beer, in which one litre of strong Bavarian beer can supply 20% of a German's energy requirements, is somewhat exaggerated, as Czechs, Danes and Belgiums also have high per capita beer consumption figures. They do not suffer harmful effects from additives any more so than do the Germans. The Germans also show no special concern over other items in their diet which could equally put them at risk.

Many countries regard it impertinent of the Germans to criticise legislation of other countries, as there is no evidence that the limits set outside Germany constitute a health risk. Germany's own legislation for foods safety is no more stringent than in other countries. Many also believe the Germans are using a tax law to to exercise import trade restrictions whilst pretending that it has something to do with food safety. It was therefore no surprise when their defence of the action brought about by the Brasseries du Pecheur was rejected in 1983. However, they continued to argue the case and it was not settled until 1987 in favour of the French brewers.

DEFENCE AND SUPPORT OF THE REINHEITSGEBOT

The Germans were firmly hardened in their resolve during the time that the case was before the European Court of Justice, as they strongly defended another similar case brought before the courts in the city of Aachen in 1984. In this case, a complaint was filed against the use of the Reinheitsgebot by the Centre for Fighting Unfair Competition. The distributor of the Belgium brewers Stella Artois had been prohibited from selling their beer in Germany because it contained 20% rice. The court would not await the result from the European Court of Justice and stated that the Reinheitsgebot did not hinder trade but protected the consumer. They stopped the beer being imported.

Like other nations the Germans are fiercely proud of their traditions especially ones that have served them well. The tradition of the Reinheitsgebot has only served them well because Germany is one of the few countries to enforce such a law and gives them a distinct trading advantage. It has also had other less obvious advantages for the German brewing industry as it has meant that the German brewer has perfected his brewing techniques to such a degree that he is able to brew without additives and this has given them a deservedly good reputation as brewers.

The Germans, despite the decision made against them are determined to carry on brewing according to the Reinheitsgebot and show other brewers that good beer can be brewed without the use of additives. It is doubtful if brewers from other countries will be able to produce a beer with the use of additives which will give them any significant marketing advantage as far as quality is concerned. Competitors will also loose out strongly on sales resistance to any product which is not brewed according to the Reinheitsgebot and face strong competition in respect of price, as the German brewers work on low profit margins and high sales volumes. The initial attack on the German market is most likely to be on the premium bottled beer market, where the price is less critical.

The criticism which the Reinheitsgebot has received over the last few years has focused attention on it and has started a wave of support outside Germany. In the U.S.A., where many German brewing traditions were taken over by immigrant German brewers, there is perhaps the greatest following. Some American brewers are marketing special brands which are publicised as being brewed in accordance with the Reinheitsgebot, with analyses carried out by the Weihenstephan Brewing Institute which certify that they are pure. A mini-brewer even formed the Reinheitsgebot Brewing Company in Texas.

In the U.K. a Cornish brewer in 1987 introduced a range of beers which are claimed to be the first to be brewed in the U.K. without the use of artificial additives, preservatives and colouring. Interestingly, the head brewer was quoted as saying that they are brewed according to the standards of the Reinheitsgebot. This is obviously an enterprising development for an English brewer and time will tell if it is the start of a trend in the U.K.

Ironically, adoption of the Reinheitsgebot by other countries would indicate to the Germans that their policy was correct though it would steal their marketing advantage and do them more harm than five years of opposition, which has strengthened their resolve and given their industry massive publicity.

OTHER PURE BEER LEGISLATION

The Isle of Man, although part of the United Kingdom, has its own Parliament called the Tynwald. One of the laws on the statute book from 1874 is a beer purity law which prohibits the use of materials other than malted barley, hops, yeast and water for brewing. However, unlike the Reinheitsgebot there is a concession allowing the use of sugar.

Most Swiss beer is also brewed to a similar purity code. Lager beers with a strength of 10.5% to 12.0%, special beers with a strength of 11.5% to 13.8% and 'Festival' beers with a strength of 12.0% to 13.5% can only be brewed to the purity code of the Swiss Brewers' Association. Other types of beer such as strong ale, diät beer and non-alcoholic beer can be brewed with adjuncts. The type of adjunct, however, has to be declared on the label.

CONCLUSION

This book has been written without any intentional bias and it is hoped that the facts collected together have been useful to the reader in helping him form his own views on the subject. The historical information was included as the author believes that facts relating to the present can convey a more complete picture when they are seen in an historical context. It is often one area of study which is undervalued and neglected with scientific and technical subjects.

It may be irresponsible to finish without giving some guidance to those who have read this book expecting either warnings or reassurance about additives, adulterants or contaminants in beer.

The author's personal interpretation of the present situation is that the health of the beer consumer is now less at risk from beer than at any time in the past and that the risks will continue to decrease as time passes. The beer consumer is no more at risk from the additives, adulterants and contaminants of beer than from those in other food products. The risks of consuming greater amounts of certain beer contaminants has in some cases increased, but the brewing industry and the government are aware of the problems and are acting to decrease the risks. Some contaminant problems, such as increasing nitrate levels, are not problems confined to beer and it is perhaps advantageous to the reader to refer to books about additives and adulterants in other foods in order to put the situation completely in perspective. Some suitable books can be found in the bibliography. Although there are areas for concern which are being monitored such as the levels of N-nitroso compounds and ATNC there is no blatant use of substances which are even suspected of being a danger to health in the light of current knowledge.

However, there is no need to be complacent. At no one time is it safe to suppose that we are at the pinnacle of knowledge on any subject or that everything possible is being done to safeguard health. As far as the general public are concerned, they are going to be the last people who will be made aware of the health risks of any dangerous substances discovered in beer. The Brewing industry protects its position for economic reasons and information on substances likely to be a danger to health is given to brewery companies by the Brewers' Society through confidential communications. Government committees other than the FAC are governed by the Official Secrets Act in order to control the passage of information. It is assumed that this is necessary in order that the general public is not needlessly brought to a state of panic by the release of information which has not been fully evaluated. If allowed to happen this could have undesirable economic, social and political consequences. Events following Edwina Curry's comments on the Salmonella contamination of eggs in 1988 illustrate how sensitive people are to the release of information on food and how accurate that information needs to be.

It may be, in view of this, that the consumer would like to make his own evaluation of the situation and decide for himself what passes his lips. One big hindrance to the consumer who wishes to make a judgement on whether he should drink a particular beer, is the fact that it does not have the ingredients listed on a label as do other foods. There seems to be no justification for withholding this information or reason to suppose it would be impractical to give it. It is very unlikely that any great inconvenience would be caused to the brewers and that they would benefit by the gain in confidence from consumers who have seen the moves to avoid product labelling as suspicious.

In some instances the numbers of additives permitted to be used for the same purpose seem to be unnecessarily large. It has been policy with some foods to restrict the numbers of permitted additives and to require more detailed specifications for those which are permitted. If this course of action were taken with brewing additives the least it would do would be to make ingredient labelling more feasible.

The other aspect of beer adulteration which affects the consumer's pocket rather than his health is dilution. The evidence for the extent of this practice is very hard to come by but what little evidence there is available, indicates that it is practised. More published evidence would be of interest to most beer drinkers.

One substance which often gets no mention when ingredients and additives are discussed is alcohol, or more specifically, ethyl alcohol. This is perhaps because it is neither one nor the other. However, it is present in beer in quite large amounts because it is produced by yeast from the main ingredients sometimes with the help of additives. It is an integral part of the product, but because it is slowly and silently produced and not added from drums or tankers, its presence is uncontrolled and often unmeasured. There is no requirement to indicate the alcoholic content of a beer and when strength is indicated it is usually the original gravity which is given. This is the strength of sugar in the wort before fermentation and only partly indicates the potential for alcohol production.

If alcohol were an additive and the desirability of its presence were decided by the toxic and harmful effects it had on the consumer, it would either be banned or its use severely restricted. The short term ill-effects of drunkenness and the ensuing hangover are well known, whereas the long term effect of damage to the nervous system, the liver, the gastro-intestinal system, the pancreas and the heart, and the effect on blood pressure are less well known. Sensible drinking limits which limit the long term ill-effects of alcohol are as low as 10.5 pints per week for men and 7 pints per week for women. The fact that there is no requirement to warn people of the dangers of alcohol in beer (let alone indicate safe levels of consumption) puts the concern over additives, adulterants and contaminants in beer and the need for ingredient labelling in a shadow.

APPENDIX 1

HERBS USED IN BREWING

ALDER, COMMON *Alnus glutinosa* (Gaertn.)
Habitat: Europe, N. Africa and W. Asia. Grows near water on marshy ground and riversides.
Description: Small tree; smooth, ovate leaves; autumn catkins.
Part used: Twigs, bark and buds.
Taste: Astringent. Contains much tannic acid.
Use: In traditional Norwegian brewing used to colour and flavour beer; considered to save malt. In Britain, in the seventeenth century, bruised bark and buds were used to flavour new ale to produce a spring tonic beer. It has the property of stopping diarrhoea.

ALECOST *Tanacetum balsamita* (Linn.)
Synonyms: Costmary. Balsam Herb. Chrysanthemum balsamita.
Habitat: Native of Asia but established in Europe in the sixteenth century.
Description: Perennial: 2–4 ft high; long, broad, upwardly pointing, grey-green leaves; creeping roots; small yellow flowers in loose clusters.
Part used: Leaves.
Taste: Balsamic. More aromatic than Tansy to which it is related.
Use: It was used in Britain up to the seventeenth century to give ale a spicy flavour, but its use declined after that. It must have been imported in the Middle Ages, as its cultivation in Europe did not become established until the sixteenth century when it must have enjoyed a period of popularity. Several sources of the time record it as being a common garden herb.

ALOES *Aloe vera* (Linn.)
Synonyms: Sea-houseleek. Sea-ay-green. *Aloe perryi* (J. G. Baker)
Habitat: India, southern and eastern Africa, Middle East and West Indies. The variety preferred by brewers in the nineteenth century came from the Island of Socotra and were known as Socotrine or Succotrine aloes. They grow best near the sea.
Description: Succulent perennials; thick, long roots from which come long, fleshy leaves with small teeth along their edges; flowers in erect terminal spikes.
Part used: Juice from the leaves which is clarified and dried to form a paste ready for use.
Taste: Very bitter with a strong flavour which can be ojectionable. The Socotrine aloes had the most preferred flavour of the several varieties available.
Use: In Britain in the eighteenth and early nineteenth centuries they were used as an illegal substitute for hops. However, their purging nature restricted their use to a partial replacement of not more than ½ oz per barrel. They were widely used for medicinal purposes.

ASH,COMMON *Fraxinus excelsior*
Habitat: Europe.
Description: Tall tree; grey bark; compound leaves; winged seeds.
Part used: Leaves and bark.
Taste: Bitter.
Use: The leaves are laxative, diuretic and promote perspiration. The German Abbess Hildegarde in her Physica Sacia in 1079 wrote: 'If thou desirest to make a beer from oats and hops boil it also with the addition of gruz and several ash leaves, as such a beer purges the stomach of the drinker and eases his chest.' It must have been used for its tonic properties.

AVENS *Geum urbanum* (Linn.)
Synonyms: Herb Bennet. Clove Root.
Habitat: Most of Europe, Russia and Central Asia. In woods and hedges.
Description: Slim perennial plant; 1–2 ft high; rhizomatous root; hairy, deep green leaves; small, yellow flowers; mass of dark crimson fruit.
Part used: Root, which has to be dried slowly to retain its aroma.
Taste: Pleasant, clove-like aroma when grown in dry, sandy soil.
Use: Recorded use between the seventeenth century and the nineteenth century in Britain and Germany, though it is likely to have been used in earlier times. It was the predominant flavouring of Augsberg beer and was also said to prevent it from turning sour. It was suspended in a bag in the cask or vat. It was also an ingredient of Brunswick Mum.

BALM *Melisa officinalis* (Linn.)
Synonym: Lemon balm.
Habitat: Southern Europe, western Asia and North Africa.
Description: Bushy shrub; wrinkled, bright green, serrated, ovate leaves; small, white flowers in bunches.
Part used: Stem and leaves.
Taste: Lemon.
Use: In Britain used before the introduction of isinglass to fine beer.

BELLADONNA *Atropa belladonna* (Linn.)
Synonyms: Deadly nightshade.
Habitat: Central and southern Europe, North Africa, south west Asia and U.S.A. Grows in waste places and prefers the shade.
Description: Perennial, bushy shrub; 2–4 ft high; thick, fleshy, white roots; dark green, oval leaves; bell shaped, red flowers; shiny black berries.
Part used: Root and leaves.
Taste: Bitter, mainly due to the alkaloid hyoscyamine.
Toxicity: Poisonous, due to the alkaloid atropine.
Use: As an adulterant in the nineteenth century to replace hops. It also gave beer intoxicating and sedative qualities.

BITTER BEAN *Ignatia amara* (Linn.) *Strychnos ignatii* (Berg.)
Synonyms: St Ignatius's Beans. Ignatius Beans. *Faba amara.*

Habitat: Philippines.
Description: A large, woody, climbing shrub; fruit is a large, pear shaped berry; seeds are 1 in. long, dull black brown, ovate, hard and horny.
Part used: Seed.
Taste: Very bitter. No odour.
Toxicity: Poisonous. Contains strychnine
Use: In Britain as a substitute for hops in the nineteenth century. Nux Vomica to which it is related was usually preferred as it was cheaper, but Morrice preferred its use as he considered it to be a wholesome substitute for hops with a pleasant, bitter taste. In a brown stout recipe, 6 lb was used with 2 cwt of hops in the copper.

BLACKBERRY *Rubus fructicosus*
Synonyms: Bramble.
Habitat: All continents. Grows in hedges, open pasture and on the edges of woodland.
Description: A prickly, rambling creeper; soft black fruit.
Part used: The berry.
Taste: Fruity and sweet when ripe.
Use: In Britain in the eighteenth and nineteenth centuries to produce Blackberry ale. The berries were added to the fermenting vessel and the beer required a short maturation of 6 weeks.

BOG BEAN *Menyanthes trifoliata* (Tournef.)
Synonyms: Buckbean. Marsh Trefoil.
Habitat: Europe. Grows in swamps, damp places, ditches and moors.
Description: Creeping plant; thick spongy rhizome; juicy, smooth leaves on long stalks; flowers in a spike on the end of a long stalk; seeds small, shiny and red. The name of the plant is derived from the fact that the leaves look like those of the broad bean.
Part used: Leaves.
Taste: Bitter, due to the glucoside menyanthin.
Use: Used in Ireland in early times to flavour beer. Use in Britain and other parts of Europe recorded in the nineteenth century. 1 oz said to replace ½ lb of hops. Used medicinally as a tonic and to prevent scurvy; in larger doses it was used to purge the digestive system and to cause vomiting.

BOG MYRTLE *Myrica gale* (Linn.)
Synonyms: Sweet Gale. Bayberry. In northern Europe also known as Pors, Porst and Porze.
Habitat: Northern Europe and Asia. Grows in bogs and on moors.
Description: Twiggy bush; sexes on different plants; grey green leaves with aromatic yellow glands; brown catkins.
Part used: Leaves, fruit and twigs.
Taste: Bitter and astringent. Odour is fragrant.
Use: In Scotland from early times to flavour beer, where it was known as Roid. In Ireland from early times where it was known as Ridiog. In England, from

Mediaeval times, first by monks, particularly in Yorkshire, to produce a popular beer known as Gale ale or Gale Beer. In Scandinavia from the fourteenth century right up until the twentieth century, either by itself or with hops or other herbs for flavour. It also made beer more intoxicating but if too much was used it could have bad after-effects and cause abortion in pregnant women.

BROOKLIME *Veronica beccabunga* (Linn.)

Habitat: Europe. Grows in shallow streams and pond edges.

Description: Creeping plant; succulent hollow stems with roots at intervals; oval leaves on rising stems; flowers in axillary racemes.

Part used: Leaves and young stem.

Taste: Bitter and unpleasantly pungent.

Use: In Britain in the eighteenth and nineteenth centuries in scurvy grass ale because of its antiscorbutic properties rather than for its flavour.

BROOM *Cytisus scoparius* (Linn.)

Synonym: Sarothamnus scoparus (Koch.).

Habitat: Central Europe and northern Asia. Grows on heaths, hillsides and sandy pasture.

Description: Shrub; 3–5 ft high; numerous long, straight and slender green branches; yellow flowers.

Part used: Tops of young branches.

Taste: Bitter.

Use: In Britain commonly in the seventeenth and eighteenth centuries for flavouring beer. It was either boiled with the wort during brewing or added by the publican as an infusion. In 1710, bittering substances other than hops were prohibited, but retailers were allowed to add infusions of broom to make broom ale. The same applied, incidentally, to wormwood.

BURDOCK *Arctium lappa* (Linn.)

Habitat: Europe and North America. It grows on waste ground and roadsides.

Description: Biennial; large, dull, pale green leaves; large tap root; tubular, purple-red flowers which produce burrs covered with hooked spines.

Part used: Root.

Taste: Strong, warm, sweet and liquorice-like.

Use: In Britain, in botanic beers as a flavouring, in the nineteenth and twentieth centuries.

BURNET, GREAT *Sanguisorba Officinalis* (Linn.)

Synonyms: Common Burnet. Garden Burnet.

Habitat: Europe. Grows in wet meadows and shady places in mountainous districts.

Description: Erect plant; 2 ft high; long, black root; pinnate leaves on long stalks; oblong heads of deep purple-brown flowers.

Part used: Whole herb.

Taste: Astringent.

Use: It has tonic properties so was used in herbal tonic beers in Britain. It was used in Brunswick Mum in the seventeenth century.

CALUMBA *Jateorhiza calumba* (Miers)
Synonym: Colombo.
Habitat: East Africa.
Description: Dioecious climbing plant; tuberous roots; small flowers.
Part used: The newer parts of the root.
Taste: Aromatic and bitter without astringency.
Toxicity: No toxic effects.
Use: In Britain in the nineteenth century as a hop substitute.

CAPSICUM *Capsicum annuum*
Synonym: Guinea Pepper.
Habitat: Originated in tropical America but now cultivated world wide.
Description: Herbaceous annual; 1 to 3 ft high; white flowers; fruit 0.5 to 11 inches long of various colours; kidney shaped seeds.
Part used: Fruit.
Taste: Pungent and hot but not the hottest of peppers.
Use: In Britain in the eighteenth and nineteenth centuries. The source of Capsicum must have been West Africa as it is commonly referred to in brewing literature of the time as Guinea pepper, as well as Capsicum. It was used to give beer a warm taste and supposedly to preserve it. A recommended rate of usage was 3 lb to 20 barrels.

CARAWAY *Carum carvi* (Linn.)
Habitat: Northern Europe, central Asia and America.
Description: 1.5 to 2 ft tall; branched stem; well spaced leaves; umbelliferous white or reddish flowers; carrot-like root; flat, greenish-brown fruit.
Part used: Seed, which technically is a fruit.
Taste: Warm and spicy. Pleasant aromatic odour.
Use: Throughout Europe from the Middle Ages. It was one of the herbs used in gruit and was mentioned as a stock item in the Gruit House of Cologne in 1393. In traditional Norwegian brewing it was used as a flavouring in the mash tun. It was used in Britain in the eighteenth and nineteenth centuries in a similar way to coriander, to flavour and to give an impression of strength to beer, but it was less popular. It was used at the rate of about 2 lb per 20 barrels.

CARDAMOMS *Elettaria cardamomum* (Maton)
Habitat: South India and Sri Lanka.
Description: Horizontally spreading stem; large leaves; racemes of yellow flowers; fleshy rhizome; oblong fruit containing small, dark red-brown seeds.
Part used: Seeds.
Taste: Powerful, pungent and aromatic. Used in curry powders.
Use: In spiced ales, notably Brunswick Mum.

CARROT, WILD *Daucus carota* (Linn.)

Habitat: Central and southern Europe, parts of Asia and America. Grows on grassland, wasteland and roadsides. Common by the sea.

Description: Finely divided leaves like the cultivated carrot; the root is thin and white with an unpleasant taste; umbelliferous flowers producing small seeds.

Part used: Seeds.

Taste: Warm, pungent and bitter. It gives beer a peach or apricot flavour.

Use: In Britain in the eighteenth and nineteenth centuries to replace some of the hops when the price was high. It was used with hops in the copper. It was also used to make an Apricot ale by suspending a bag of the seed in the finished beer for three weeks. It is antiscorbutic, so acted as a tonic in the beers to which it was added.

CASSIA *Cinnamomum Cassia* (Blume)

Synonyms: Chinese Cinnamon. Bastard Cinnamon.

Habitat: Wild in Assam. Cultivated in the Far East, Mexico and South America.

Description: Tropical evergreen tree; fruit the size of an olive.

Part used: Bark and the small unripe fruit known as buds.

Taste: Warm, more pungent, less sweet and less delicately flavoured than cinnamon to which it is related.

Use: In Britain in the nineteenth century to give a pungency to badly managed beer. It was added to the cask. It was not widely used compared with cinnamon.

CATECHU, BLACK *Acacia catechu* (Willd.)

Synonym: Cake catechu.

Habitat: Burma, India.

Description: Tree.

Part used: The wood taken when the tree has reached 1 ft diameter, from which an extract containing tannic acid is prepared.

Taste: Astringent with sweet after-taste.

Use: In Britain in the nineteenth century to precipitate protein material in the beer. It is said that it was thrown out of solution and did not appear in the finished product. It may also have been used with hop substitutes which were recommended to be used with tannin material because they were naturally deficient in it.

CENTAURY *Erythraea Centaurium*

Habitat: Central and southern Europe, North Africa and Asia minor.

Description: Small, erect plant; about 6 in. high; pale green, pointed leaves; white and red flowers; woody root.

Part used: All the plant except the root.

Taste: Bitter.

Use: In British herb beers around the seventeenth century for its tonic properties. In Britain in the eighteenth century as a bittering herb to partially replace hops. It was boiled in the copper with the hops.

CHAMOMILE, COMMON *Anthemis nobilis* (Linn.)
Habitat: Europe, North Africa and Asia minor.
Description: Creeping plant; tufts of leaves and flowers; the nature of the leaves give the plant a feathery look; daisy-like flowers.
Part used: The whole herb except roots.
Taste: Bitter with an aromatic fragrance.
Use: In Britain in the nineteenth century recommended as a wholesome hop substitute. One pound was equivalent to five pounds of hops. It was also used as an ingredient of herb beers having tonic and aperient properties.

CHASTE TREE *Vitax agnus castus*
Habitat: Mediterranean area and Asia Minor.
Description: Tree; young shoots covered with a grey down; fingered foliage; flower spikes of pale lavender; fruit like peppercorns.
Part used: Fruit and twigs.
Taste: Warm and peculiar, said to be much like cannabis.
Use: By the Cauchi to flavour beer before the introduction of hops. It is a male anaphrodisiac used by Athenian matrons in he sacred rites of Ceres for securing chastity.

CHERRY LAUREL *Prunus laurocerasus* (Linn.)
Habitat: Asia minor and southern Europe.
Description: Small, evergreen tree; broad, glossy, oval leaves; blackish bark; flowers in white racemes; fruit like black cherries.
Part used: Leaves.
Taste: Bitter with aroma of hydrocyanic acid.
Toxicity: Poisonous and intoxicating.
Use: In Britain in the nineteenth century, as a hop substitute and to improve intoxicating properties. A dangerous adulterant.

CHIRETTA *Swertia chirata* (Buch-Ham.)
See **GENTIAN.**

CINNAMON *Cinnamomum Zeylanicum* (Nees.)
Habitat: Sri Lanka but now cultivated in many eastern countries. It is best grown in sand in a sheltered spot.
Description: Tropical, evergreen tree; 20 to 30 ft tall; young shoots speckled a greeny-orange; leathery leaves; small white flowers.
Part used: The dried inner bark of shoots.
Taste: Warm, sweet and aromatic.
Use: In Mediaeval times in Britain and Germany in spiced ale. Used in scurvy grass ale in the eighteenth and nineteenth centuries. Also known to have been sometimes used in porter.

CLARY, COMMON *Salvia sclarea* (Linn.)
Habitat: Native of southern Europe. Grows on rocky banks and cliffs, in stony soils and on wasteland and roadsides.

Description: Biennial; 2 to 3 ft high; pairs of heart-shaped leaves as large as a hand; blue flowers in long loose terminal spike.

Part used: Syrup extracted from the plant.

Taste: Aromatic, warm and bitter.

Toxicity: Narcotic.

Use: In Britain in the eighteenth and nineteenth centuries as a hop substitute. It also gave beer intoxicating properties described by one writer as 'an effect of insane exhilaration of the spirits, succeeded by severe headache'. Syrup of clary fermented with a little yeast was advocated in an eighteenth century recipe for reviving beer.

CLOVES *Eugenia caryophyllata* (Thumb.)

Habitat: Molucca Islands and southern Philippines.

Description: Small, evergreen tree; bright-green leaves; flowers in bunches at the ends of the branches.

Part used: The undeveloped flowers.

Taste: Distinctive and aromatic.

Use: In Mediaeval Germany in spiced beer. In Britain in the eighteenth century, included with sugar in a recipe for a solution added to bottled beer in order to feed it. The same solution was also used to recover tap droppings. Perhaps used for its slightly germicidal effect.

COCCULUS INDICUS *Anamirta paniculata* (Colebr.)

Synonyms: India berry. Occulus Indicus. Fish berry.

Habitat: India and Sri Lanka.

Description: A climbing plant; corky bark; heart-shaped leaves; pendulous flowers; browny-black fruit about the size of a black current, but wrinkled.

Part used: The berry.

Taste: Bitter.

Toxicity: Poisonous narcotic. Gives severe headache, vertigo, distressing sickness and if enough taken can cause death.

Use: In Britain used in porters and stouts as an illegal bittering agent. It was used at the rate of 4 lb per 20 barrels in the copper. It was also used outside brewing to catch fish. It was thrown into the water and when the fish ate it, they became intoxicated and floated on the surface.

COLOCYNTH *Citrullus colocynthis* (Schrad.)

Synonym: Bitter apple.

Habitat: Native of Turkey but also found in Asia and Africa.

Description: An annual plant resembling the water melon; globular fruit the size of an orange containing a white spongy pulp in which are brown seeds.

Part used: The fruit pulp.

Taste: Bitter and nauseous.

Toxicity: Violent purgative and poisonous.

Use: In Britain in the nineteenth century as a hop substitute. Its nature did not deem that it could ever have been very popular and it is only referred to as a substance which was experimented with, but it is one of the substances which public analysts tested for.

COLTSFOOT *Tussilago farfara* (Linn.)

Synonyms: Clayweed. Coughwort.

Habitat: Europe, Asia, North Africa and introduced into North America. Grows in poor soil on waste ground.

Description: A yellow composite flower is produced on a stem before the appearance of leaves which are hoof-shaped, 4 in. across and arranged in a rosette.

Part used: Leaves.

Taste: Slightly bitter. Expectorant.

Use: In Britain to make a traditional herb beer called cleats. Derived its name clayweed because it grows on clay soils. A wine made from it is also called clayt wine. The herb is a very good remedy for coughs so the beer is most likely to have been a medicinal beer.

CORIANDER *Coriandrum sativum* (Linn.)

Habitat: Cultured in Britain but a native of southern Europe, the Middle East and Africa.

Description: A slender stemmed plant; 1 to 3 ft high; shiny palmate leaves; umbelliferous pinky-white flowers; thick, round, red-yellow seeds.

Part used: Seeds. They have to be dried to obtain a fragrant taste.

Taste: Hot, spicy and aromatic. Used in curry powders.

Toxicity: If too much used the seeds can become narcotic.

Use: In Britain in the eighteenth and nineteenth centuries. It was sometimes used by small brewers to make the beer taste stronger than it was. One pound of seeds was said to be equivalent to one bushel of malt. It was also used to flavour various types of beer either by adding it to the copper or by putting it in a bag suspended in the cask. Used in the copper for this purpose the normal rate was about 1 lb per 10 barrels.

COWSLIP *Primula veris* (Linn.)

Habitat: Europe. Grows in dry meadows and woodland.

Description: A spring flowering plant; leaves a little similar to the primrose, near the ground and in a rosette; flowers in an umbel on the end of a long stalk.

Part used: Flowers and root.

Taste: Acrid with an odour of anise.

Use: In Britain in the eighteenth century to make cowslip ale; the flowers were added to ale in the cask and left for two weeks before bottling. The rate of use was one bushel of flowers to a barrel of ale. The powdered roots were also added to the copper to produce an ale which helped cure giddiness, nervous trouble and wakefulness. Cowslips are slightly narcotic and have a sedative value.

CUMIN *Cuminum cyminum* (Linn.)

Habitat: Indigenous to Egypt but grown in Europe, the Middle East, India and China.

Description: Slender branched stem; about 1 ft high; leaves divided into segments; small pink or white flowers; oblong, small, flattened fruit called seeds.

Taste: Warm, slightly bitter and aromatic but less pleasant than caraway seeds.

Use: It was a common spice in Mediaeval Europe and was used in spiced beers but lost favour to caraway. Its cultivation and use in beer was recorded as far north as Norway.

DANDELION *Taraxacum officinale* (Webber)

Habitat: Europe, North Africa and Asia. Grows on pastures, wasteland, verges and almost any place.

DANDELION

Description: Thick brown tap root; rosette of toothed leaves close to the ground; yellow flowers on stalks.
Part used: Leaves.
Taste: Bitter and appetitive.
Use: In Britain its use has been recorded from the seventeenth century and even into the twentieth century. It was popularly used in herbal or botanic beers. It was often used with other herbs such as nettle and burdock, the later being a well known combination. Dandelion beer was also known of in Canada.

DARNEL, BEARDED *Lolium temulentum* (Linn.)
Synonyms: Cockle. Cheat. Ray-grass.
Habitat: Europe. It used to be a very common weed in corn fields.
Description: A grass with flat rough wide leaves; flowers in a spike; seeds with long awns.
Part used: Seed.
Taste: Noxious and acrid.
Toxicity: Narcotic producing intoxication. In the sixteenth century Tabernaemontanus recorded it as being a dangerous herb when used in ale as it could cause blindness.
Use: Its use has been recorded in England, Wales and Norway between the sixteenth and nineteenth centuries. It gave beer strength and increased its intoxicating properties, which could save on the cost of malt and hops, hence the name cheats. Since it was a naturally occurring weed its properties were first discovered by accident but its use as an adulterant became quite deliberate.

ELDER *Sambacus nigra* (Linn.)
Habitat: Europe and parts of Asia. Grows on roadside verges and riversides.
Description: A small tree; creamy-white flowers; drooping bunches of small purple-black berries.
Part used: Berries and flowers.
Taste: Fruity but the flowers can produce a cat like aroma if used in too great a concentration.
Use: In Britain, in the nineteenth century, the fruit were used to colour and flavour porter. A strong, matured beer called Ebulum was produced from the fruit which was added during brewing to the copper. A pale Ebulum was also produced from unripe berries and pale malt.

ELECAMPANE *Inula helenium* (Linn.)
Habitat: Southern Europe, parts of Asia and North America. It grows on damp ground and in shady places.
Description: A large plant; 4 to 5 ft high; large leaves covered in down; large yellow flowers resembling the sun flower; root is large and succulent.
Part used: Root.
Taste: Slightly bitter with an agreeable camphoraceous aroma.
Use: It was put into new beer in Britain to give it medicinal properties. The herb was well known as a tonic and expectorant. Culpeper also said that in ale it

'cleareth, strengtheneth and quickeneth the sight of the eyes.' It was also used in an English version of Brunswick Mum.

EYEBRIGHT *Euphrasia officinalis* (Linn.)

Habitat: Europe. Grows on heaths, dry pasture and chalky soil.

Description: Small, annual plant; 2 to 8 in. high; deeply cut leaves; small, white or pale purple flowers variegated with yellow.

Part used: The plant above the root.

Taste: Bitter and acrid.

Use: In Britain in the seventeenth century to make a herb beer. It was used to cure weakness of the eyes and even believed that it could restore sight. Beer was one of several ways in which it was administered.

FLAX *Linum usitatissimum* (Linn.)

Synonym: Linseed.

Habitat: All temperate and tropical regions.

Description: An erect annual; 1 to 2 ft high; smooth, solitary stems; leaves about 1 in. long; blue flowers; round fruit about the size of a pea containing 10 brown seeds.

Part used: Seed.

Taste: Mucilaginous and slightly unpleasant.

Use: The envelope or testa of the seed contains 15% mucilage and this was said to have been used to supplement the extract in the brewing of porter in the nineteenth century. The cake produced after the linseed oil had been extracted was the part which was probably used as the oil would have had a detrimental effect on the head of the porter. The oil has medicinal properties but it was not these which were being sought.

FOXGLOVE *Digitalis purpurea* (Linn.)

Habitat: Europe. Grows in woodland clearings, on heaths, roadsides and places where there is little soil.

Description: A tall plant; several stems 3 to 4 ft high; purple bell-shaped flowers in a spike on the end of the stem.

Part used: Leaves.

Taste: Bitter and nauseous.

Toxicity: Poisonous, violent purgative producing languor, giddiness and even death.

Use: In Britain as an adulterant in the nineteenth century in order to give beer bitterness and intoxicating properties.

GALANGAL *Alpinia officinarum* (Hance)

Synonym: China root.

Habitat: China and Java.

Description: The plant grows to about 5 ft high; long, narrow leaves; flowers have white petals with red veins and are arranged in a terminal spike; root is a rhizome which is 2 to 3 in. long and ¾ in. thick.

Part used: Root.

Taste: A mild ginger flavour.

Use: It has been used in Europe as a spice for many centuries and was used in some spiced ales. It was used in the preparation of China ale in the eighteenth and nineteenth centuries by suspending it in a bag with coriander seed in a cask of ale. It was also used in Purl. A spirituous extract of the root was known as Spirit of Maranta and had an oily nature. If added to beer in cask or bottle it floated on the surface and prevented oxidation of the beer.

GENTIAN, YELLOW *Gentian lutea* (Linn.)

Synonym: Bitter wort.

Habitat: The mountains of western and central Europe.

Description: Thick, hollow stems; about 2 to 4 ft tall; large, stiff, oblong leaves; large yellow flowers in upper leaf axils; long, thick, fleshy roots.

Part used: Root.

Taste: Penetrating bitterness.

Use: It was used in Britain in the nineteenth century as a hop substitute. 1 lb was used to replace 7 lb of hops. It was also used in recipes for scurvy grass ale and purl. Chiretta (or Chirata) which is the dried tissue of the Green Gentian of Northern India was also used in the same period but 1 lb was used to replace 10 lb of hops.

GERMANDER, SAGE LEAVED *Teucrium scorodonia* (Linn.)

Synonyms: Wood sage. Wood Germander.

Habitat: Western Europe. It grows in woody and hilly places amongst bushes and under hedges.

Description: A perennial; 1 ft tall; grey-green, oblong leaves; spikes of yellow and purple flowers; creeping roots.

Part used: The tops of the plant.

Taste: Bitter and said to resemble the flavour of hops.

Use: In Britain it was used at one time to fine beer but if too much was used it made the beer dark. In Jersey it was called Ambriose and was used there and in some other districts as a substitute for hops.

GINGER *Zingiber officinale* (Rosc.)

Habitat: Native of Asia but cultivated elsewhere in the world such as Africa and the West Indies.

Description: A perennial root which creeps underground sending up shoots in the springtime; long, narrow leaves; yellow or white flowers on a flowering stalk.

Part used: Root.

Taste: Hot and spicy.

Use: It has been widely used in Britain in the eighteenth and nineteenth centuries in brewing. It has been used as a flavouring in the copper at the rate of around 3 lb per 20 barrels, as an ingredient in cures for tainted beer, as an ingredient in cures for slow fermentations and as an ingredient used to assist in the cleansing of beers. It is also of course the main ingredient of Ginger Beer.

GRAINS OF PARADISE *Aframomum melegueta*
Synonyms: Melegueta pepper. Guinea grains.
Habitat: West Africa.
Description: Perennial plant producing small, dark aromatic seeds.
Part used: Seed.
Taste: Hot, astringent and pungent. It has been used as a substitute for true pepper from the fifteenth century. There has also been a trade in them as cardamom substitutes as they are also known as false cardamoms.
Use: It was widely imported into Britain in the eighteenth and nineteenth centuries, though today its use is confined to Africa. It was added to beer to give a warm flavour and to give the impression of strength. It could be added to the copper at the rate of 3 lb per 20 barrels but tended to remain in the beer and make it cloudy. It was therefore often added in the tun after being ground.

GROUND IVY *Glechoma Hederacea* (Linn.)
Synonyms: Ale hoof. Tun hoof. Gill ale.
Habitat: Central Europe. Grows in shady spots on moist land.
Description: A perennial creeping plant; puts out roots from the stem at intervals; two dark-green leaves which resemble true ivy at every joint on the stem; flowers are a violet-blue in the leaf axils.
Part used: Leaves.
Taste: Bitter and aromatic with a balsamic aroma.
Use: It was used from Mediaeval times probably until the eighteenth century for fining beer especially in country areas.

HENBANE *Hyoscyamus niger* (Linn.)
Synonym: Hogsbean.
Habitat: Central and southern Europe, Asia and North America. Grows on waste ground and roadsides.
Description: Grows to about 2 ft; coarsely digitate leaves, which like the stem are covered in hairs; yellow flowers with purple veins.
Part used: Leaves.
Taste: The plant has a nauseous odour which on drying disappears and a bitter flavour predominates.
Toxicity: Contains the poisonous alkaloids, scapolamine and hyoscyamine. It causes intoxication, delirium, nausea, vomiting, feverishness and death.
Use: In Britain as an adulterant in the nineteenth century to make beer more intoxicating and as a hop substitute. It was also known to have been used in Norway in Mediaeval times but was banned because of its poisonous nature.

HOLY THISTLE *Carbenia benedicta* (Berul.)
Synonyms: Blessed Thistle. *Cnicus benedictus* (Gaetn.). *Carduus benedictus* (Steud.).
Habitat: Southern Europe and Asia Minor. Cultivated in Britain. It grows in waste stony places.
Description: A slender red stem growing to 2 ft; long, narrow leaves with wavy

margins ending in spines; pale-yellow flowers in green, prickly heads.
Part used: Leaves and flowering tops.
Taste: Bitter.
Use: In British herb beers around the seventeenth century for its tonic properties. In Britain in the eighteenth century as a hop substitute. It was boiled in the wort for 15 minutes.

HOREHOUND, WHITE *Marrubium vulgare* (Linn.)
Habitat: Southern and central Europe, North Africa and the Near East. Grows on wasteland and roadside verges.
Description: A bushy plant with several stems; about 1 ft high; wrinkled hairy leaves; small white flowers in leaf axils.
Part used: Leaves.
Taste: Bitter, slightly salty, spicy aroma and expectorant.
Use: In Britain used to flavour beer before the introduction of hops and afterwards used as a hop substitute. A special horehound beer was also produced which was known for its medicinal and tonic properties. The herb was said to have the property of making stale beer taste new. It was also involved in a fraud with old hops, which after washing and drying were sprinkled with horehound juice in order to revive their bitterness and green colour.

HORSERADISH *Cochleaia amoracia* (Linn.)
Habitat: Europe. Grows alongside ditches and rivers.
Description: Large cylindrical root; shiny ovate leaves 1 to 2 ft long; small, white flowers.
Part used: Root.
Taste: Hot, spicy and biting.
Use: In Scurvy Grass Ale, Purl and other spiced ales in the eighteenth century. It was added to the finished beer rather than in the copper, as it looses its bite if boiled.

HYSSOP *Hyssopus officinalis* (Linn.)
Habitat: Asia and southern Europe. It likes mountainous regions.
Description: Small, evergreen bush; about 1 to 2 ft high; varieties have either blue, pink or white flowers on the end of the stem.
Part used: Flowers and leaves.
Taste: Aromatic and astringent.
Use: In Britain in the seventeenth century it was used to make hyssop ale which was a medicinal beer. The herb has expectorant and carminative properties. The herb was also used to fine beer and was included as an ingredient in a cure for ropey beer in the eighteenth century.

IRIS *Iris germanica, Iris pallida* and *Iris Florentina*
Synonyms: Iris germanica is known as the German Iris or Flag Iris which is cultivated in gardens in Britain.
Habitat: Mediterranian region and northern India.
Description: I. germanica has sword-like leaves and flower stems which are 2 to

3 ft long; large deep-blue or purple-blue flowers; thick creeping fleshy rhizomes.

Part used: The root. The root of the three above-mentioned species is known as Orris root and is collected in Italy.

Taste: The dried root has a slightly bitter taste and an aroma resembling violets.

Use: It was used in Britain in the eighteenth century in a cure for sour beer and to flavour beer by hanging it in a muslin bag suspended in a cask.

JALAP *Convolvulus jalapa* (Linn.)

Synonym: Jalap Bindweed.

Habitat: South America and Mexico. Cultivated in Europe.
Description: Climbing plant; tuberous root; crimson flowers.
Part used: Root. Sold whole or sliced.
Taste: Nauseous but sweetish bitterness.
Use: In Britain in the nineteenth century. It was used to excite fermentation with the purpose of preventing beer turning sour, as sometimes happened with languid fermentations. It was described as counteracting the 'binding tendency' of the other ingredients. The rate of usage was 3 oz to 20 barrels. Its use must have been highly valued as the Burton brewers, when defending their position over the use of adulterants in 1829, admitted that it was one of the few ingredients, other than malt or hops, which they used.

JUNIPER *Juniperus communis* (Linn.)
Habitat: Europe, North Africa, North America and northern Asia. It grows in dry rocky soils and on moorland.
Description: An evergreen shrub; 4 to 6ft high; needle-shaped leaves; berries which turn bluish-black in the autumn of its third year.
Part used: Though juniper oil is distilled from the berries an extract used in brewing was prepared by boiling the branches for two hours. Sturdy, mature branches bearing berries were preferred. Branches with too many needles gave too much resin.
Taste: Plant has an aromatic scent. The berries are slightly bitter-sweet, fragrant and spicy.
Use: In traditional beers of Norway and Sweden. The above extract was poured over the malt during mashing or juniper twigs were used to provide a strainer in the mash tun. A Juniper beer was also produced without the use of malt. Juniper was also used in Germany and Britain. In Britain it was an ingredient of Purl and was sometimes added to Porter as a preservative.

LEMON *Citrus limonum* (Risso.)
Habitat: Northern India but cultivated in the Mediterranian area.
Description: A small irregularly branched tree; oval, green leaves; flowers on stems in leaf axils; familiar ovoid, yellow fruit.
Part used: Rind of fruit.
Taste. Rind is bitter with familiar aromatic aroma.
Use: In Britain, in the eighteenth and nineetenth centuries, to flavour and help preserve beer. It was an ingredient of China Ale.

LING *Calluna vulgaris*
Synonym: Scots heather.
Habitat: Europe. Grows on moorland.
Description: Rambling plant; needle shaped leaves; small, pink flowers rich in nectar.
Part used: Flowers.
Taste: Astringent.
Use: In England, Scotland, Ireland and Scandinavia from early times in the production of Heather ale.

LIQUORICE *Glycrrhiza glabra* (Linn.)

Habitat: Southern Europe and Asia. Introduced to Britain 400 to 500 years ago.

Description: Perennial; 1 ft tall; rhizomes just under the ground; delicate foliage with leaflets on a midrib; small, pale-blue flowers (sometimes purple-blue or yellow-white) like pea flowers.

Part used: The rhizome, which should be the thickness of the little finger, ruddy outside and yellow inside.

Taste: Sweet, pleasant and soothing.

Use: In Britain in the nineteenth century. It was used as colouring at the rate of ½ lb per barrel in some ales, porters and stouts. It was also used to supplement the unattenuated sugar left after fermentation in some beers. Spanish juice was the name given to liquorice root boiled in water and then evaporated to concentrate it. It was drawn out into golden strings which set solid. It was in this form that it was used in many recipes.

MADDER *Rubia tinctorum* (Linn.)

Habitat: Southern and mid Europe.

Description: Tall plant with stems about 8 ft, but so weak that they are often on the ground; stems prickly; leaves have spines on underside of midrib; flowers in loose yellow spikes.

Part used. The root.

Taste: Astringent and slight odour.

Use: The powdered root when added to beer gives it a red colour. Used as a colouring with Red Saunders and Walnut in the recipe for Brunswick Mum.

MARJORAM, SWEET *Origanum marjorana* (Linn.)

Habitat: Portugal, but cultivated elsewhere in Europe.

Description: Height 12 to 18 in.; mossy stem; hairy leaves on which are oil glands; small, white flowers in spikes on the end of the stems.

Part used: Leaves.

Taste: Aromatic and pleasant.

Use: As a flavouring in Brunswick Mum.

MARSH ROSEMARY *Ledum palustre*

Synonyms: Wild Rosemary. Marsh Cistus. Porsch.

Habitat: Northern Europe, Asia and North America. Grows in damp woods and marshes.

Description: Perennial; 2 ft high; woody stem; narrow, oblong, pointed leaves which contain a volatile oil; flowers blue or variegated white and purple.

Part used: Leaves.

Taste: Spicy.

Toxicity: Narcotic, poisonous after-effects.

Use: One of the main components of gruit. It has the effect of making beer more intoxicating.

MEADOWSWEET *Spiraea ulmaria* (Linn.)

Synonym: Meadsweet.

Habitat: Mid and northern Europe. Grows in meadows and on moist banks.

Description: Fern-like foliage; tufts of delicate, creamy-white flowers; 2 to 4 ft high; thick rhizome.

Part used: Flowers.

Taste: Flowers have an almond-like fragrance but the leaves are tart and bitter.

Use: In Mediaeval Britain the flowers were used in beer and wine, most likely for medicinal purposes. It was one of fifty ingredients in a drink called Save mentioned in Chaucer's Knights Tale.

MELILOT *Melilotus arvensis* (Lamk.)

Habitat: Not a native of Britain, but in the sixteenth century it was widely cultivated as a fodder crop. It spread to the wild and became a common weed which spoilt many fields of corn.

Description: Perennial; 2 to 4 ft high; branched, erect stem; trifoliate leaves; racemes of yellow or white flowers; black, one-seeded pods.

Taste: Sweet aroma which on drying becomes stronger. This is due to coumarin which is present in hay and woodruff. The taste, however, is pungent and bitter. Malt contaminated with melilot produces a beer which is quite nauseous.

Effect: It was described in the eighteenth century as 'being apt to cause fevers, colics and other distempers in the body'.

MINT, GARDEN *Mentha viridis* (Linn.)

Habitat: Native of the Mediterranean region.

Description: Square stem; 2 ft high; short-stalked leaves; small, pink or lilac flowers in the axils of upper leaves; creeping roots.

Part used: Leaves.

Use: In British herb beers around the seventeenth century for its tonic properties.

MUSTARD, WHITE AND BLACK *Brassica alba* and *Brassica nigra*

Habitat: White grows in Europe. Black grows farther south to include North Africa and Asia Minor. Naturalised in the Americas.

Description: White is an erect annual; 1 ft high; large yellow cruciferous flowers. Black is a larger plant but smaller flowers. Both produce seeds in pods, though they differ in appearance.

Part used: Seed used bruised or powdered.

Taste: Hot but only have an aroma when made into a paste.

Use: In Britain in the eighteenth and nineteenth centuries used for removing taints from beers caused by infections and burnt malt. It was suspended in a bag in the beer with other ingredients such as alum and ginger. It was also used in an eighteenth century cure for a slow fermentation. An onion dipped in mustard was supposed to accelerate the fermentation.

MYRTLE *Myrtus communis*

Habitat: Southern Europe. Can just survive the British winter.

Description: Large shrub; white flowers; black flask-shaped berries.

Part used: Berries.

Taste: The dried fruit were used as an alternative to pepper corns.

Use: In Mediaeval Germany. St Hildegarde is recorded as having produced myrtle beer though this may have been beer made from bog myrtle which was used.

NETTLE, COMMON *Urtica dioica* (Linn.)

Habitat: Europe, Asia, South Africa and Australia.

Description: Heart-shaped, stinging leaves; clusters of green flowers; creeping root system.

Part used: Leaves and stem. Nettle tops are picked before June as after that they become laxative and loose flavour.

Taste: Tart and slightly salty. Nettles contain formic and lactic acid.

Use: In Britain to make a tonic beer. It is considered a pleasant drink which will cure rheumatic and gout pains.

NUTMEG *Myristica fragrans* (Houtt.)

Habitat: Moluccas and other islands of the East Indian Archipelago.

Description: Tree; 25 ft high; smooth, grey-green bark; glossy, aromatic evergreen leaves; fleshy fruit like an apricot. Inside the fruit is a scarlet net-like membrane called the mace, enclosing a dark brown shell inside which is the seed or nutmeg. The mace and the nutmeg are dried to produce two different spices.

Part used: The nutmeg and the mace (to a lesser extent) were both used in beer.

Taste: Nutmeg is sweet, warm and aromatic. Mace is also warm with a nutmeg-like aroma and a sharp, fatty taste.

Toxicity: Nutmegs contain a toxic substance called myristicin which in excess can cause harmful effects on the liver.

Use: In spiced beers in Europe since Mediaeval times.

NUX VOMICA *Strychnos nux-vomica* (Linn.)

Habitat: Egypt and Sri Lanka.

Description: Medium-sized tree; short, crooked trunk; smooth bark; green-white, funnel-shaped flowers; fruit the size of a large apple with a hard orange rind; inside the fruit are five flat round seeds, 1 in. across and ¼ in. thick.

Part used: Seed.

Taste: Bitter but very little aroma.

Toxicity: Poisonous. Contains the alkaloids strychnine and brucine.

Use: In Britain, in the nineteenth century, it was an adulterant used as a hop substitute. When brewers' druggists were investigated in 1818, many admitted to selling *Cocculus Indicus* when in fact they were using Nux vomica which is more poisonous.

OAK GALLS *Quercus infectoria*

Habitat: Asia Minor, Syria and Iran. The best galls were imported from Aleppo in Turkey.

Nut

Cross section of nut

seed

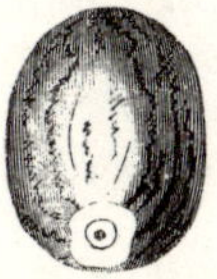

NUTMEG

Description: Galls are excrescences produced by the attack of the female Gall Wasp piercing under the bark of young shoots. Good Aleppo galls are about ¾ in. in diameter, a dark-bluish-green or olive-green with no perforations made by the escape of the insect. *Q. infectoria* on which the galls are found is a small type of oak, indigenous to the area, seldom attaining 6 ft in height.

Taste: Bitter and astringent. They contain gallotannic acid.

Use: In Britain, in the nineteenth century, in order to fine and colour beer. A recipe for finings required 3 lb of nut galls and 4 oz of potash to be boiled for three hours, which was then made up to 12 lb with water. When cooled 2 pt of spirit of wine was added and after it had settled it was bottled for use.

ORANGE, BITTER *Citrus vulgaris* (Risso.)
ORANGE, SWEET *Citrus aurantium* (Linn.)

Habitat: India, China and southern Europe.

Description: Small tree; oval, evergreen leaves; round, orange fruit.

Part used: Rind of fruit.

Taste: Bitter.

Use: In Britain, in the nineteenth century, used as a hop substitute or in attempts to improve the flavour in certain types of beer. Sometimes orange peas or Curacoa oranges, the unripe fruit of the *Citrus aurantium*, were used instead. They are dark-grey-brown.

PENNYROYAL *Mentha pulegium* (Linn.)

Habitat: Parts of Europe and Asia.

Description: Weak, prostrate stems, 3 to 12 in. long; grey-green, hairy leaves; flowers in whorled clusters.

Part used: The herb.

Taste: A type of mint with a strong aromatic aroma.

Use: In British herb beers around the seventeenth century for its tonic properties. In Brunswick Mum. In Kykeon, a beer described by Homer.

PEPPER, LONG *Piper officinarum*

Habitat: India and the Far East.

Description: Climbing vine; dark-grey fruit.

Part used: Fruit.

Taste: Weakly aromatic but pungent taste.

Use: It is not now used in the Western world as *Piper nigrum* is preferred, but between the sixteenth and eighteenth centuries it was used in brewing in Britain. It was added to the copper at the rate of about 3 lb per 20 barrels to give the beer spirit and to save on malt. Two pods of pepper in a bottle of beer were said to stop excessive working which made the contents fly. Its use in beer in Norway has also been recorded.

POLYPODY, COMMON *Polypodium vulgare* (Linn.)

Synonyms: Polypody of Oak. Oak Fern.

Habitat: Europe. It grows in sheltered places such as on the roots and stumps of trees and on stone walls.

Description: A creeping rhizome; thick, woody fern fronds hanging in tresses.
Part used: Root and sometimes the leaves.
Taste: Expectorant.
Use: In Scurvy Grass Ale for its tonic and medicinal properties.

POPPY, WHITE *Papaver somniferum* (Linn.)
Synonym: Opium Poppy.
Habitat: Indigenous to Asia Minor but cultivated in other Eastern countries.
Description: Erect, herbaceous annual; flowers vary in colour from white to reddish-purple; seeds in capsules, the walls of which are filled with a kind of latex.
Part used: Seed capsules.
Toxicity: Poisonous narcotic.
Use: In Britain, in the nineteenth century, used as an adulterant in order to make beer more intoxicating. Seizures were recorded in 1827. Multum which was a mixture of opium and other substances was a common way in which it was used. At 5/- to 6/- per pound in 1818 it was cheaper than *Cocculus indicus* at $1\frac{1}{2}$ guineas per lb.

QUASSIA *Quassia excelsa*
Habitat: West Indies and America.
Description: A tree growing to between 50 and 100 ft; smooth, grey bark; oblong leaves; small, yellow-green flowers.
Part used: Bark or wood from the root or trunk in the form of chips.
Taste: Intensely bitter, but it leaves an unpleasant after-taste in the mouth for a long time afterwards.
Toxicity: Slightly narcotic but has tonic properties and was prescribed by physicians at the rate of ¼ oz per day.
Use: As a hop substitute, in Britain, in the nineteenth century. It was considered to be one of the purest and least harmful of the many adulterants which were used for this purpose. One pound was said to be equivalent to 16 lb of hops, but care was needed in its use, because if over-used it could cause unpleasant flavours. It was used in the curing of alcoholics when quassia and sulphuric acid were used to destroy their appetite for drink.

Quassia amara, which is a smaller shrub, has a more intense bitterness but was not used so much as *Q. excelsa*.

QUILLAIA *Quillaja saponaria* (Molina.)
Synonym: Soap tree.
Habitat: Chile and Peru. Cultivated in India.
Description: Tree growing to 50 to 60 ft; oval, shiny leaves; terminal, white flowers; tough, thick, dark bark.
Part used: The bark.
Taste: Acrid and astringent.
Toxicity: Contains Saponin which is poisonous in large amounts and causes gastric irritation in smaller amounts.

Use: It has been used for over 100 years in soft drinks in the U.K. and the U.S.A. for its foam forming properties. It was used in beer in the nineteenth century but this is now not permitted. It was advertised for sale as Eccumen, Foaming Powder or Frothing Powder and in its production was often mixed with magnesia to give it bulk and sulphites as a preservative.

RED SAUNDERS *Pterocarpus santalinus*
Synonym: Red Sandlewood.
Habitat: India and Sri Lanka.
Description: Tree; 25 to 30 ft tall; rough bark; spikes of yellow flowers.
Part used: The wood.
Taste: Slightly astringent.
*Use:*It was used as a colouring in some British beers. It occurs in the recipe for Brunswick Mum with Madder and Walnut.

RHUBARB, TURKEY *Rheum palmatum*
Habitat: China and Tibet. It is called Turkey Rhubarb because it was imported through Turkey.
Description: Large leaves; brown stem, 6 to 10 ft long; thick root. Large, compared to Garden Rhubarb.
Part used: The root.
Taste: Astringent, nauseous with a characteristic odour.
Use: It is a purgative and tonic and was added to scurvy grass ale because of these properties. The rhubarb referred to in recipes would not have been Garden Rhubarb, *Rheum rhaponticum* (Willd.) as it was only introduced to Britain in 1777. It is much milder and was grown from some seeds purchased from Russia by Hayward, an apothecary from Banbury, who sold it in competition with Turkey Rhubarb.

ROSEMARY *Rosmarinus officinalis* (Linn.)
Habitat: Europe and Japan.
Description: Shrub; 6 ft high; leathery, evergreen leaves; small pale-blue flowers.
Part used: The tops of flowering shoots which are considered better than the leaves alone.
Taste: Bitter with a camphor-like aroma.
Use: It was added to some beers from Mediaeval times until about the eighteenth century, most likely for its tonic properties. It was either boiled in the copper or added to the cask after fermentation.

SAGE, COMMON *Salvia officinalis* (Linn.)
Synonym: Garden sage.
Habitat: Southern Europe but cultivated in Northern Europe.
Description: Shrubby perennial; about 1 ft high; square stem covered in down; oval, evergreen leaves; small, purple flowers.
Part used: Leaves.
Taste: Bitter.

Use: It was used for many centuries to produce Sage ale which was a type of some note. It was a tonic beer and Gerard said of it: 'No man need to doubt of the wholesomeness of sage ale, being brewed as it should be with Sage, Betony, Scabious, Spikenard, Squinnette and Fennell Seed.' It was added to the copper, or two or three drops of extract were added to a cask of beer.

ST JOHN'S WORT *Hypericum perferatum* (Linn.)
Habitat: Europe and Asia. Grows on roadsides and in woods and meadows.
Description: Perennial, growing to about 2 ft; stems with branched tops; leaves with transparent spots on them; yellow flowers.
Part used: All the plant except the root.
Taste: Astringent and aromatic.
Use: In traditional Norwegian beers, from Mediaeval times, to make them more intoxicating. It was used in the mash tun. The herb also has expectorant properties so may have been used for medicinal reasons as well.

SASSAFRAS *Sassafras officinale* (Lees and Eberm.)
Habitat: Native of the east coast of North America, but can be cultivated in Britain.
Description: Deciduous tree; 20 to 40 ft high; clusters of green-yellow flowers; dark-blue fruit.
Part used: Oil extracted from the root.
Taste: Pleasant, aromatic and slightly spicy.
Use: As a flavouring in some types of botanic beers, particularly in America. Later, commonly used in soft drinks. The young shoots were used in Virginia for making a kind of beer. In Britain in the eighteenth century it was included in a recipe for Brunswick Mum.

SCURVY GRASS, COMMON *Cochlearia officinalis* (Linn.)
Habitat: Originated in northern Europe. Prefers a salty environment and grows on the coast.
Description: Small ground-hugging plant; heart-shaped, fleshy leaves on long stalks and higher elongated leaves without stalks; small, white flowers in bunches on the end of the stems.
Part used: Fresh, lower leaves before flowering.
Taste: Bitter and salty. Sharp pungent aroma.
Use: In Britain in the eighteenth and nineteenth centuries in Scurvy Grass Ale.

SENNA *Cassia acutifolia* (Dell.)
Habitat: Middle East.
Description: Small shrub; 2 ft high; long, spreading branches; leaflets in 4 or 5 pairs; small, yellow flowers; seeds in pods.
Part used: The leaves and the pods the latter being the milder.
Taste: Nauseating, sweet and mucilaginous.
Use. In Britain, in Scurvy Grass Ale, for its purgative qualities.

SWEET FLAG *Acorus calamus* (Linn.)
Synonym: Calamus. Sweet Sedge.
Habitat: Europe, U.S.A. and parts of Asia. Grows near to water.
Description: Reed-like plant; sword-shaped, yellow-green leaves; small, yellow-green flowers in a spike; rhizomatous roots.
Part used: Rhizome.
Taste: Slightly bitter, warm, spicy, pleasantly aromatic.
Use: In Britain in the eighteenth and nineteenth centuries as a hop substitute. One pound of root was said to be equivalent to 6 lb of hops. It was sliced and boiled in the copper with the hops. It was also used in some beers for the sake of its own flavour and was said to give spirit to the beer. Besides being boiled with the wort it was also used by suspending it in a linen bag in a cask of beer. Used in the production of Purl in this way.

TANSY *Tanacetum vulgare* (Linn.)
Habitat: Europe. Grows in hedge bottoms and waste places.
Description: Perennial; 2 to 3 ft high; leaves covered with blisters which contain an aromatic oil; clusters of dull-yellow, button-like flowers.
Part used: Leaves and flowers.
Taste: Bitter and spicy with a camphor-like aroma.
Toxicity: It can be dangerous if used in large amounts and should not be used during pregnancy.
Use: In Britain in the seventeenth century it was used for making Tansy Ale. It was also used as a flavouring in traditional Norwegian beers by putting it in the mash tun.

THYME, GARDEN *Thymus vulgaris* (Linn.)
Habitat: Southern Europe, North Africa and Asia Minor. Cultivated in Britain.
Description: Hard, branched stems; 4 to 8 in. high; small, grey-green leaves; woody, fibrous root; purple flowers in whorls on the ends of the stems.
Part used: Leaves.
Taste: Tart and spicy taste, with a pleasant aroma.
Use: An ingredient in Brunswick Mum.

TOBACCO *Nicotiana tabacum* (Linn.)
Habitat: Most sub tropical countries.
Description: Erect stem; 3 to 6 ft high; numerous large, ovate, pale-green leaves; long fibrous roots.
Part used: Leaf.
Toxicity: Contains the alkaloid nicotine, which is a virulent poison affecting the digestive and circulatory organs.
Use: In Britain in the nineteenth century as an adulterant to replace a portion of the hops. It was also used in Norway but reported that it made people sick when they drank the beer.

VERVAIN *Verbina officinalis* (Linn.)
Habitat: Europe, China and Japan.

Description: Perennial; ribbed, erect stem which is loosely branched and sparsely leafed; small pale lilac flowers.
Part used: Leaves.
Taste: Tart and slightly astringent.
Use: In Britain, in the eighteenth century, with hyssop to fine beer. It supplemented the action of the hyssop which in too large a quantity could give beer a tang.

WALNUT *Juglans nigra* (Linn.)
Habitat: Europe, the Middle East and India.
Description: Tree; 40 to 60 ft high; fruit is the familiar nut.
Part used: Leaves.
Taste: Bitter with a pleasant aroma.
Use: In Brunswick Mum probably as a colouring agent. The leaves produce a brown dye when boiled.

WATERCRESS *Nastutium officinale*
Habitat: Europe and Asia. Grows in water.
Description: Creeping stem; brown-green leaves; small white flowers.
Part used: Leaves and stem.
Taste: Slightly bitter.
Use: In scurvy grass ale because of its antiscorbutic properties.

WOOD BETONY *Stachys betonica* (Benth.)
Synonym: Wild Hop. This refers to the appearance of the flower rather than any other properties.
Habitat: Europe. Grows in woods and amongst shrubby growth on moorland.
Description: Perennial; 1 to 2 ft high; thick, woody root; thin stem; leaves in pairs, but more at the base of the stem; spike of purple flowers.
Part used: Whole herb.
Taste: Bitter, aromatic, astringent and said to taste somewhat like tea.
Use: In Britain, from Mediaeval times until the nineteenth century, for making Betony Ale which was said to be an ale of note. As a herb it was regarded as a general cure-all particularly for headaches. Chewing the fresh leaves before drinking was said to prevent drunkenness.

WOODRUFF *Asperula odorata* (Linn.)
Habitat: Europe and parts of Asia. Grows in woodland and shady places.
Description: Perennial; 6 to 8 in. high; whorls of dark-green leaves at intervals on the stem; small, white flowers at the tip.
Part used: Sprigs and essence.
Taste. Aroma of coumarin which is found in hay. It also fixes other aromas.
Use: In Germany essence is added at serving to Berliner-Weisse beer. German law forbids it to be added during brewing.

WORMWOOD *Artemisia absinthium* (Linn.)
Synonym: Green Ginger.
Habitat: Europe. Grows on poor soils.

Description: Branched stem about 2 ft high covered with grey hairs; downy, segmented leaves; yellow, globular flowers.

Part used: Leaves.

Taste: Bitter and strongly aromatic.

Toxicity: In strong doses it is addictive and poisonous.

Use: In Britain to make wormwood ale. Mentioned by Pepys in 1660. In 1710, wormwood and broome, by law, were the only herbs which could be used in beer besides hops. In the eighteenth century a physician said of wormwood beer that it ' . . . does not only hurt but by degrees weakens the natural heat of the stomach and sends heavy dull vapours to the head and so prejudiceth the eyes'. In the early nineteenth century it was used as a hop substitute, but it was said that by 1830 its use had been discontinued. Some brewers preferred Sea Wormwood (*Artemisia maritima*) which is less bitter. It was also a traditional herb used in brewing in Norway.

YARROW *Achillea millefolium* (Linn.)

Synonym: Milfoil.

Habitat: Europe, North America and mid Asia. It grows on wasteland and pasture.

Description: Perennial; creeping root system; rough stem; feathery looking leaves; small, daisy-like flowers.

Part used: Leaves when young.

Taste: Bitter and tart with a spicy scent.

Toxicity: In excess it can cause dizziness, ringing in the ears and even madness. It causes especially bad hangovers.

Uses: It has been used in many northern countries including Iceland, Norway, Sweden, Germany and Britain to brew beer. It was one of the principal herbs used in gruit. It was considered that it made beer more intoxicating as well as contributing flavour.

YEW *Taxus baccata*

Habitat: Europe, North Africa and western Asia.

Description: Tree; 40 to 50 ft high; red-brown, peeling bark; small, dark-green leaves; bright-red fruit.

Part used: Tips of branches.

Toxicity: Poisonous.

Use: In Britain in the nineteenth century as an adulterant in order to make beer more intoxicating.

APPENDIX 2

LEGISLATION IN THE NINETEENTH CENTURY IN THE U.K.

1800 An Act was passed which allowed the temporary use of sugar for nine months.

1802 No materials other than malt, hops and water allowed for brewing.

1811 to 1817 Solutions of burnt sugar allowed for colouring porter only. In Ireland the use of colouring was not stopped until 1827.

1812 An Act passed again allowing the temporary use of sugar.

1830 Beer duty repealed. Unmalted grain prohibited on brewery premises.

1842 The use of roasted or black malt recognised by law. Regulations imposed in respect of its manufacture, sale and storage at breweries in 1856 and 1867.

1847 Sugar allowed for use in brewing beer but not molasses or sugar on which customs duty had not been paid.

1855 The keeping of raw grain on brewery premises was prohibited and plain rollers had to be used to crush the malt so that they could not be used for raw grain.

1860 Adulterations of Food Act introduced.

1862 Brewers of spruce or black beer were exempted from the licence duty levied on brewers, on provision that they used no hops, bittering substances or yeast in their beer.

1862 Hop duty repealed and the use of substitutes allowed. It was extended to Ireland in 1864.

1865 The practice of using treacle with glucose in beer prohibited by the Inland Revenue, as treacle was a prohibited material.

1870 Brewers had to use a sugar stock book and enter sugar used in the brewing book.

1872 Adulteration of Food Act introduced. It was an amendment of the 1860 Act.

1874 Any sugar substance was permitted for use in beer or in the manufacture of colouring.

1875 Sale of Food and Drugs Act introduced.

1880 Present system of beer duty imposed. The brewer was allowed to brew with whatever materials he pleased providing they were not deleterious.

1885 Deleterious substances specifically prohibited in beer in the Inland Revenue Act.

1888 Saccharine prohibited in beer in order to protect the revenue.

APPENDIX 3

CURRENT LEGISLATION IN THE U.K.

The Food Act 1984
Antioxidants in Food Regulations 1978
The Preservatives in Food Regulations 1979
The Emulsifiers and Stabilisers in Food Regulations 1980
The Miscellaneous Additives in Food Regulations 1980
Colouring matter in Food Regulations 1973
Lead in Food Regulations 1979
Arsenic in Food Regulations 1959
Materials and Articles in Contact with Food Regulations 1978
The Food Labelling Regulations 1984
Alcoholic Liquor Duties Act 1979
The Beer Regulations 1985

APPENDIX 4

PERMITTED PRESERVATIVES IN THE U.K.

Sulphur dioxide E220
Sodium sulphite E221
Sodium hydrogen sulphite E222
Sodium metabisulphite E223
Potassium metabisulphite E224

Calcium sulphite E226
Calcium metabisulphite E227
Benzoic acid E210
Sodium benzoate E211
Potassium benzoate E212
Calcium benzoate E213
Methyl 4-hydroxybenzoate E218
Methyl 4-hydroxybenzoate sodium salt E219
Ethyl 4-hydroxybenzoate E214
Ethyl 4-hydroxybenzoate sodium salt E215
Propyl 4-hydroxybenzoate E216
Propyl 4-hydroxybenzoate sodium salt E217

APPENDIX 5

ADDITIVES IN COMMON USE AT THE END OF THE NINETEENTH CENTURY

HOP SUBSTITUTES
Quassia, catechu, chiretta, calumba root, gentian, tannic acid, resins, camomile and calyso.
STIMULATING MATTER
Grains of paradise, Guinea pepper.
FININGS AND HEADINGS
Isinglass, fishy matter(sole skins).
Quillaia bark, saponin extracts.
PRESERVATIVES AND YEAST PURIFIERS
Sulphurous acid, potassium metabisulphite, potassium sulphite, calcium metabisulphite, calcium bisulphite, calcium sulphite, magnesium bisulphite, sodium metabisulphite, sodium sulphite, ammonium sulphite.
Salicylic acid.
Formalin.
NEUTRALISERS FOR ACID BEER
Potash, soda ash.
YEAST FOODS
Sodium phosphate, potassium phosphate, calcium phosphate, magnesium phosphate, potassium sulphate, magnesium sulphate, magnesium chloride, sodium chloride.
COLOUR
Caramelised cane sugar, caramelised starch sugar, caramelised malt or grain extract, caramel containing liquorice.
LIQUOR TREATMENT
Calcium chloride, sodium chloride , calcium sulphate, magnesium sulphate.

APPENDIX 6

ADJUNCTS IN USE AT THE END OF THE NINETEENTH CENTURY

CEREALS

Raw maize, flaked maize, gelatinised maize, malted maize, flaked maize flour, torrified flaked maize, maize grits.

Raw rice grain, desiccated rice, flaked rice, gelatinised rice, malted rice, rice grits.

Flaked barley.

Oats.

Farina.

Wheat.

Sago.

GLUCOSE AND SACCHARUMS

Glucose from sago, maize, potato and barley.

Various so-called saccharins or saccharums were also produced from the above.

SUGARS

Molasses, raw sugar, cane sugar, invert sugar, honey, cane sugar with liquorice, malt extracts.

APPENDIX 7

ADULTERATION OF BEER IN THE NINETEENTH CENTURY

Year ended 31 March	Number of Samples Total taken	Positive	Adulterants found
1844	36	30	Not described.
1845	7	0	Not described.
1846	9	1	Not described.
1847	41	25	Not described.
1848	4	2	Not described.
1849	3	3	Not described.
1850	5	1	Not described.
1851	55	47	Not described.
1852	15	9	Not described.
1853	24	9	Not described.
1854	10	8	Not described.

1855	59	54	Not described.
1856	15	9	Not described.
1857			No report.
1858	21	–	No details available.
1859	21	10	Grains of paradise; ground chicory.
1860	12	8	Grains of paradise; ground chicory; Sweet flag.
1861	42	29	Grains of paradise; sweet flag; *Cocculus indicus*; ginger; liquorice; capsicum; carraway; coriander.
1862	5	4	Not described.
1863	4	4	Grains of paradise; *Cocculus indicus*.
1864	26	20	Grains of paradise; *Cocculus indicus*; tobacco; capsicum; iron sulphate.
1865	31	23	Grains of paradise; coriander; mustard; sweet flag; iron sulphate.
1866	13	10	Drugs; tobacco.
1867	17	15	Grains of paradise; treacle; liquorice.
1868	34	30	Grains of paradise; treacle; caramel; sugar.
1869			No report.
1870-1	19	14	Grains of paradise; treacle; liquorice; ginger; unmalted oats.
1872	134	25	Grains of paradise; liquorice; sugar; caramel; salt; copperas.
1873	18	6	Grains of paradise; liquorice; sugar; caramel.
1874	60	8	Liquorice; caramel; flour; salt.
1875	59	28	Liquorice; caramel; treacle; salt; ginger; tobacco.
1876	38	18	Sugar; caramel; treacle.
1877	89	61	Not described.
1878	75	40	Grains of paradise; liquorice; sugar; glucose.
1879	50	27	Liquorice; caramel; sugar.
1880	673	526	Not described but principally water.
1881	212	155	Principally water.
1882	89	63	Sugar; treacle.
1883	82	–	No details available.
1884	33	–	No details available.
1885	48	–	No details available.
1886	106	–	No details available.
1887	1,999	624	Water; sugar.
1888	2,630	450	Water; sugar.
1889	2,027	329	Water; sugar.
1890	948	275	Water; sugar.
1891	1,956	459	Water; sugar.

1892	2,044	392	Water; sugar.
1893	3,026	620	Water; sugar; saccharine.
1894	2,820	498	Water; sugar; saccharine.
1895	2,838	507	Water; sugar.
1896	3,462	534	Water; sugar.
1897	5,013	790	Water; sugar.

From 1887 all adulterated samples were taken from public houses. Water and sugar were not prohibited ingredients of beer but as they had been added after the beer had left the brewery in order to defraud the customer and H.M. Customs and Excise they were regarded as adulterants.

APPENDIX 8

PERCENTAGES OF ADJUNCTS USED IN BREWERS' GRISTS CALCULATED FROM H.M. CUSTOMS AND EXCISE STATISTICS

In the statistics published from 1847 until 1896, sugar was the only adjunct identified separately. In 1897 when cereal adjuncts started to become more popular, after being allowed for use in brewing in 1880, they were also identified separately. Percentages have been calculated from the weights used rather than the 'bushel equivalents of malt' used which were shown in some of the earlier statistics. The 'bushel equivalent' values were changed several times in the early statistics which tended to confuse matters when the usings were seen together. However, 'bushel equivalents' were originally used by the Excise to calculate whether a brewer was producing the full amount of wort from his materials and were not created for statistical purposes. If a brewer produces below a permitted minimum, a materials charge is made and duty is paid on the minimum permitted amount, even though he has not obtained it. Since 1880, 2 cwt of sugar has been equivalent to 1 quarter of malt (3 cwt) but even by 1900 the expected extracts from the two materials would have made a figure of 2.5 cwt more realistic.

For these reasons the percentage calculations have been based on weight, with the assumption that although there has been a change in the amount of extract which has been obtained from various materials, these have generally been increases which have roughly kept in line with each other as technological advances have been made.

PERCENTAGE SUGAR USED IN THE TOTAL GRIST 1847 TO 1896

1847	0.66%	1857	0.14%	1867	2.01%	1877	3.44%	1887	6.93%
1848	0.22%	1858	0.25%	1868	1.82%	1878	4.79%	1888	7.24%
1849	0.15%	1859	0.18%	1869	2.00%	1879	4.81%	1889	8.14%
1850	0.09%	1860	0.56%	1870	1.23%	1880	5.93%	1890	8.67%
1851	0.05%	1861	0.59%	1871	1.48%	1881	5.54%	1891	8.82%
1852	0.05%	1862	0.52%	1872	1.25%	1882	5.76%	1892	9.59%
1853	0.06%	1863	0.55%	1873	2.56%	1883	5.50%	1893	9.22%
1854	0.26%	1864	0.29%	1874	3.40%	1884	5.50%	1894	9.64%
1855	0.15%	1865	0.26%	1875	3.92%	1885	6.10%	1895	9.85%
1856	0.12%	1866	0.48%	1876	3.44%	1886	6.40%	1896	10.07%

PERCENTAGE ADJUNCTS BY WEIGHT FROM 1897

YEAR	UNMALTED CORN	OTHER CEREAL ADJUNCTS	SUGAR	TOTAL
1897	0.39	3.04	10.46	13.89
1898	–	3.66	10.95	15.01
1899	0.43	4.30	11.64	16.37
1900	–	4.68	11.60	16.68
1910	0.16	5.63	12.68	18.67
1915	0.19	6.00	13.02	19.21
1920	0.19	5.40	11.27	16.86
1925	0.17	6.37	13.08	19.62
1930	0.20	5.35	12.88	18.43
1935	0.10	5.51	15.29	20.90
1939	0.08	5.82	15.74	21.64
1943	0.31	9.55	10.85	20.66
1945	1.78	9.65	12.93	24.36
1950	0.53	4.16	11.81	16.50
1955	0.44	4.47	14.30	19.21
1960	0.08	4.92	14.18	19.18
1965	0.46	5.56	13.70	13.70
1970	1.60	5.25	15.03	15.03
1975	4.27	3.66	15.11	15.11
1980	3.44	4.56	13.59	13.59
1985	2.58	5.88	14.27	14.27

YEAR ENDS: Up to 1851, 5 January of succeeding year.
Up to 1854, 10 October.
Up to 1969, 30 September.
Up to 1989, 31 March.

From 1971 flaked barley is under unmalted corn.

APPENDIX 9

ADDITIVES AND PROCESSING AIDS USED IN BREWING AND MALTING IN 1989 IN THE U.K.

'E' Numbers in parentheses. PA = Processing aid. A = Additive

MALTING		
Acaricides	Prevention of barley spoilage	PA
Fungicides	Prevention of barley spoilage	PA
Insecticides	Prevention of barley spoilage	PA
Calcium hydroxide (526)	Adjustment of steep liquor pH	A
Sodium carbonate (500)	Adjustment of steep liquor pH	A
Sodium hydroxide (524)	Adjustment of steep liquor pH	A
Calcium hypochlorite	Steep liquor bactericide	A
Potassium bromate (924)	Reduction of malting loss	A
Sodium bromate	Reduction of malting loss	A
Sulphur dioxide (E220)	Reduction of malting loss	A
Sulphuric acid (513)	Reduction of malting loss	A
Phosphoric acid (E338)	Reduction of malting loss	A
Hydrochloric acid (507)	Reduction of malting loss	A
Citric acid (E330)	Reduction of malting loss	A
Acetic acid (E260)	Reduction of malting loss	A
Lactic acid (E270)	Reduction of malting loss	A
Gibberellic acid	Accelerates germination	A
Hydrogen peroxide	Breaking of dormancy	A
Sulphur dioxide (E220)	Stops germination in the kiln	A
Ammonia	Prevents 'magpie' malt	A
LIQUOR TREATMENT		
Ion exchange resins	Deionisation of water	PA
Carbon	Filtration	PA
Calcium hydroxide (526)	Softening	A
Lactic acid (E207)	Softening and pH reduction	A
Phosphoric acid (E338)	Softening and pH reduction	A
Sulphuric acid (513)	Softening and pH reduction	A
Hydrochloric acid (507)	Softening and pH reduction	A
Calcium sulphate (524)	Flavour adjustment and pH reduction	A
Magnesium sulphate (518)	Flavour adjustment and pH reduction	A
Calcium chloride (509)	Flavour adjustment	A
Potassium chloride (508)	Flavour adjustment	A
Sodium chloride	Flavour adjustment	A
Sodium sulphate (514)	Flavour adjustment (rare)	A
Potassium bisulphate	Flavour adjustment (rare)	A
Chlorine	Sterilant	A
Sodium hypochlorite(925)	Sterilant	A
Chlorine dioxide	Sterilant	A

ANTIOXIDANTS		
L-ascorbic acid (E300)		A
Sodium L-ascorbate (E301)		A
Sulphur dioxide and sulphites (See Preservatives)		
COLOURING MATERIALS		
Caramel	Adjustment of beer colour	A
ENZYME PREPARATIONS		
Bromelain	Haze prevention	A
Papain	Haze prevention	A
Pepsin	Haze prevention	A
Microbial amyloglucosidase	Improvement in wort fermentability	A
Microbial α-amylase	Starch conversion and improvement in wort fermentability	A
Microbial β-glucanase	Reduction of wort viscosity	A
Microbial proteinases	Increase of amino nitrogen in wort	A
Microbial pullulanase	Improvement of wort fermentability	A
YEAST NUTRIENTS		
Ammonium chloride (510)		A
Ammonium phosphate (450a)		A
Ammonium sulphate (520)		A
B-complex vitamins(375)		A
Cupric sulphate		A
Ferrous ammonium sulphate		A
Ferrous sulphate		A
Glutamic acid		A
Yeast autolysates		A
Zinc chloride		A
Zinc sulphate		A
CLARIFYING AGENTS		
Carrageen (Irish moss)	Copper and auxiliary fining	PA
Carrageenan (E407)	Copper and auxiliary fining	PA
Sodium alginate (E401)	Copper and auxiliary fining	PA
Silicic acid preparation	Auxiliary fining	PA
Agar agar (E406)	Copper fining	PA
Furcellaran (E408)	Copper fining	PA
Acacia (E414)	Copper fining	PA
Isinglass	Yeast fining	PA
BEER STABILISERS		
Tannic acid (181)	Precipitates proteins	PA
PVPP	Adsorbs polyphenols	PA
Bentonite	Adsorbs proteins	PA

Silica hydrogel	Adsorbs proteins	PA
F0AM STABILISERS		
Propane-1,2-diol alginate (E405)		A
Acacia gum (E414)		A
ANTIFOAMS		
Dimethylpolysiloxane (900)		PA
Glyceryl monoricinoleate (E471)		PA
Glyceryl monostearate (E471)		PA
Sorbitan monolaurate (E493)		PA
Sorbitan monostearate (E491)		PA
Sorbitan monooleate (E494)		PA
Sorbitan polyoxyethylene monostearate (E435)		PA
FILTER MEDIA		
Kieselguhr		PA
Perlite		PA
Silica hydrogel		PA
Cellulose		PA
Asbestos		PA
Bentonite		PA
YEAST WASHING AGENTS		
Bisulphites (see preservatives)		PA
Lactic acid (E270)		PA
Phosphoric acid (E338)		PA
Sulphuric acid (513)		PA
Tartaric (E334)		PA
Acidified ammonium persulphate		PA
PRESERVATIVES		
See Appendix 4		

The criterion for denoting a substance an additive or a processing aid is that if there is a specific part of the brewing process which is used to remove the substance, then it is a processing aid and not an additive.

It could be argued for instance, that bromates used in malting have been shown not to survive the brewing process, so should be regarded as processing aids. However, their fate during the brewing process has only been discovered recently and it is incidental good fortune that they do not survive, rather than being something that the brewer deliberately attempts to do. On the other hand the brewer makes deliberate attempts to remove some substances which are classified as processing aids, but as they are sometimes not completely removed they would be more accurately classified as additives. The distinction between additives and processing aids may therefore not be entirely clean cut.

Some substances such as salts used in liquor treatment and hormones like gibberellic acid are naturally present in the brewing materials used and it is impossible to distinguish the natural from the added substance. Some people do not regard these substances as additives for this reason.

The list of additives may appear large but it must be remembered that a brewer may not use additives in every group and when he does, he will most likely use only one out of the group. The list is so large because the FAC, who advised on the use of suitable additives for brewing, were persuaded by the brewers and manufacturers that a wide range of substances was needed to meet different needs. Criticism was levelled at the FAC for not being critical enough when evaluating whether there was sufficient need for certain additives.

BIBLIOGRAPHY

BOOKS PUBLISHED BEFORE 1900

Anon, *The London and Country Brewer*, London, 4th ed. 1742.
Anon, *The Young Brewer's Monitor*, London, 1824.
Accum, F., *A Treatise on the Art of Brewing*, London, 1820.
Baker, G., *An Essay, Concerning the Cause of the Endemial Colic of Devonshire*, London, 1767.
Baring-Gould, Rev. S., *Devon*, London, 1899.
Bickerdyke, J., *The Curiosities of Ale and Beer*, London, 1886.
Black, W., *A Practical Treatise on Brewing*, London, 1844.
Booth, D., *The Art of Brewing*, London, 1829-34.
Brande, W., *Town and Country Brewery Book*, London, 1829.
Bryn, M. L., *The Complete Practical Brewer*, Philadelphia, 1876.
Child, S., *Every Man His Own Brewer*, London, 1790.
Death, J., *The Beer of the Bible*, London, 1887.
Donovan, M., *Domestic Economy*, London, 1830.
Faulkner, F., *Theory and Practice of Modern Brewing*, London, 1884.
Greenwood, J., *The Seven Curses of London*, London, 1869.
Hughes, E., *Treatise on Brewing of Beer and Porter*, Uxbridge, 1796.
Laurent, A., *Dictionary of the Art of Brewing*, Liverpool, 1881. (Translated, edited and enlarged by Wilson, H. M.)
Leifchild, J., *Complete Directions for Brewing*, London, 1802.
Levesque, J., *The Art of Brewing*, London, 1853.
Loveless, R. D., *Every-day Wrinkles for Brewers*, Birmingham, 1881.
Macnish, R., *The Anatomy of Drunkenness*, Glasgow, 1832.
Morewood, S., *Inebriating Liquors*, London, 1824.
Morrice, A., *Practical Treatise on Brewing*, London, 8th ed. 1834.
Scammel, G. and Colyer, F., *Breweries and Maltings*, London, 1880.
Siebel, J. E., (Editor), *One Hundred Years of Brewing*, Chicago and New York, 1903.
Southby, E. R., *A Systematic Handbook of Practical Brewing*, London, 1885.
Shannon, R., *A Practical Treatise on Brewing, Distilling and Rectification*, London, 1805.
Thausing, J. E., *The Theory and Practice of the Preparation of Malt and the fabrication of Beer*, London and Philadelphia, 1882.
Threale, T., *The Complete Family Brewer*, London, 1802.
Tizard, W. L., *Theory and Practice of Brewing*, London, 1857.
Tryon,T., *A New Art Of Brewing Beer*, London, 1691.

BOOKS PUBLISHED AFTER 1900

BREWING HISTORY

Arnold, J. P., *Origin and History of Beer and Brewing*, Chicago, 1911.
Corran, H. S., *A History of Brewing*, Newton Abbot, 1975.

Filby, F. A., *A History of Food Adulteration and Analysis*, London, 1934.
Hackwood, F. W., *Inns, Ales and Drinking Customs of Old England*, London, 1909.
King, F. A., *Beer Has a History*, London, 1947.
Lucas, A. and Harris, J. R., *Ancient Egyptian Materials and Industries*, London, 1962.
Mathias, P., *The Brewing Industry in England 1700- 1830*, Cambridge, 1959.
Monckton, H. A., *A History of English Ale and Beer*, London, 1966.
Monckton, H. A., *A History of the English Public house*, London, 1969.
Nordland, O., *Brewing and Beer Traditions in Norway*, Oslo, 1969.
Salzman, L. F., *English Industries in the Middle Ages, London*, 1923.
Wilson, C. A., *Food and Drink in Britain*, London, 1973.
Wilson, R. G., *Greene King: a business and family history*, London, 1983.

BREWING SCIENCE AND TECHNOLOGY
Baker, J. L., *The Brewing Industry*, London, 1905.
Broderick, H. M., (Editor), *The Practical Brewer*, Wisconsin, 1977.
De Clerck, J., *A Textbook of Brewing*, London, 1957.
Hough, J. S., Briggs D. E. and Stevens R., *Malting and Brewing Sciences*, London, 1971.
Lloyd-Hind, H., *Brewing Science and Practice*, London, 1938.
Pollock, J. R. A., (Editor), *Brewing Science*, London, 1979.

HERBS AND SPICES
Culpeper, N., *Culpeper's Complete Herbal and English Physician*, London, 1652.
Genders, R., *The Complete book of Herbs and Herb Growing*, London, 1980.
Gerard, J., *The Herball or Generall Historie of plants*, London, 1597.
Grieve, M., *A Modern Herbal*, London, 1931.
Keble Martin, W., *The Concise British Flora in Colour*, London, 1965
Mc Lean, T., *Medieval English Gardens*, London, 1981.
Purseglove, J. W. et al., *Spices*, London and New York, 1981.
Rosengarten, F., *The Book of Spices*, New York, 1969.

ADDITIVES
Davis, S. and Stewart, A., *Nutritional Medicine*, London, 1987.
Hanssen, M. with Marsden, J., *The New E for Additives*, Wellingborough, 1987.
Lipske, M. et al., (Center for Science in the Public Interest), *Chemical Additives in Booze*, Washington D C, 1982.
Mervyn, L., *Thorsons Complete Guide to Vitamins and Minerals*, Wellingborough, 1986.
Millstone, E., *Food Additives – Taking the Lid Off What We Really Eat*, London, 1986.
Millstone, E. with Abraham J., *Additives: a guide for everyone*, London, 1988.

ADULTERATION
London Food Commission, *Food Adulteration and How to Beat It*, London, 1988.

ARTICLES AND REPORTS

ADJUNCTS

Board of Customs and Excise, *Annual report*, Nos 1 to 71.

Brewing Correspondent, *Brewing and Distilling International*, 1976, **6**, 20.

Brewing Correspondant, *Brewing and Distilling International*, 1983, **13**, 22.

Britnell, J., *Technical Quarterly of the Master Brewers' Association of the Americas*, 1973, **10**, 176.

Canales, A. M. & Sierra, J. A., *Technical Quarterly of the Master Brewers' Association of the Americas*, 1976, **13**, 114.

Collier, J. A., *Brewing Room Book 1976-7*, 76th Issue, Pauls & White Ltd., page 63.

Collier, J. A., *Brewing and Distilling International*, 1986, **16**, 15.

Commissioners of Inland Revenue, *Annual Report*, Nos 24 and 43.

Imrie, F. K. E. & Martin, F.J., *Brewers' Guild Journal*, 1969, **55**, 252.

Lloyd, W. J. W., *Brewing Room Book 1972-3*, Pauls & White Ltd., 74th Issue, page 13.

Lloyd, W. J. W., *Journal Institute of Brewing*, 1986, **92**, 336.

Maiden, A. M., *Brewing Room Book 1978-9*, Pauls and White, 77th Issue, Page 64.

Maiden, A. M., *Brewing and Distilling International*, 1985, **5**, 36.

Marchbanks, C., *Brewing and Distilling International*, 1987, **17**, 16.

Maul, A. P. & Greenshields, R. N., *Process Biochemistry*, 1970, **5**, 39.

Roberts, R. H., & Rainbow, C., *The Brewer*, 1971, **57**, 112.

Spence, J. A., *The Brewer*, 1971, **57**, 402.

Thirlwell, A. J. & Busby, R. L., Proceedings of the Australia and New Zealand Section 17th Convention. *Journal of Institute of Brewing*, 1982, **88**, 77.

ANTIFOAMS

British Patent 1290444, 1972.

Button, A. H. & Wren, J. J., *Journal Institute of Brewing*, 1972, **78**, 443.

Evans, J. I., *Process Biochemistry*, 1972, **7**, 29.

Pierpoint, D. J., *The Brewer*, 1988, **74**, 290.

ASBESTOS

Maurer, C. H. & Coors, J. H., *Brewers' Digest*, 1975, 54-56; 58; 60-64.

Shackleton, R. & Glynn, I., *The Brewer*, 1983, **69**, 100.

COLOURING

Anon., *Caramel – an Update*, British Caramel Manufacturers' Association, 1988.

Buckee, G. K. & Bailey, T. P., *Journal Institute of Brewing*, 1978, **84**, 158.

F.A.C., *Final Report on the Review of Colouring Matter in Food Regulations 1973*, H.M.S.O., 1987.

M.A.F.F., Food Surveillance Paper No. 19, *Survey of Colour Usage in Food*, H.M.S.O., 1987.

Greenshields, R. N., *Process Biochemistry*, 1973, **8**, 17.

COPPER FININGS

Matthews, A. J. D., *The Brewer*, 1986, **72**, 384.
Rowsell, J. T., *Brewers Guild Journal*, 1966, **52**, 289.
Wylie, A., *Journal of the Royal Society of Health*, 1973, **93**, 309.

DILUTION

Howell, G. C., *The Brewer*, 1984, **70**, 237.
O'Neill, P., *What's Brewing*, Nov 1987.
The Institute of Trading Standards Administration, *The Licensed Trade – Results of a Survey*, December 1987.

ENZYMES

Butcher, D. M., *The Brewer*, 1987, **73**, 61.
Dickenson, C. J., *The Brewer*, 1987, **73**, 55.
Harding, G., *The Brewer*, 1981, **67**, 365.
Woodward, D. J., *Brewers' Guardian*, 1972, **101**, 43.

FOAM STABILISERS

Evans, J. I., *Process Biochemistry*, 1972, **7**, 29.
Winter, W. G. S., *Brewing and Distilling International*, 1971, **1**, 49.
Wylie, A., *Journal of the Royal Society of Health*, 1973, **93**, 309.

FILTER MEDIA

Ross, G. R., *The Brewer*, 1982, **68**, 52.

FININGS

Anon., *The Brewer*, 1984, **70**, 512.
Courts, A., *Brewers' Guardian*, 1971, **100**, 57.
McCreath, A. A., *Brewing and Distilling International*, 1982, **12**, 27.
Rowsell, J. T., *Brewers' Guild Journal*, 1966, **54**, 278.

FLOURIDES

Narziss, L., et al., *Brauwelt*, 1977, **117**, 1667.
Sherlock, J. C., *Journal of the Royal Society of Health*, 1984, **104**, 34.

FORMALDEHYDE

Brewing Research Foundation, Annual Report, *Journal Institute of Brewing*, 1987, **93**, 277.
Marchbanks, C., *Brewing and Distilling International*, 1986, **16**, 31.

HOP PRODUCTS

Clarke, B. J., *Journal Institute of Brewing*, 1986, **92**, 123.
Forster, A., *Brauwissenschaft*, 1979, **32**, 57.
Forster, A., *Forum der Brauerei*, 1983, **36**, 21.

LEAD AND ARSENIC

Crosby, N. T., *The Analyst*, 1977, **102**, 225.
Department of the Environment, Pollution Paper No. 12, *Lead in Drinking Water*, H.M.S.O. 1977.
Department of the Environment, Pollution Paper No. 19, *Lead in the Environment*, H.M.S.O., 1983.

M.A.F.F. Food Surveillance Paper No. 8, *Survey of Arsenic in Food*, H.M.S.O., 1982.

M.A.F.F. Food Surveillance Paper No. 10, *Survey of Lead in Food: Second Supplementary Report*, H.M.S.O., 1982

Madge, A. G. M., *Journal of the Royal Society of Health*, 1987, **107**, 72.

Pohlmann, R., *Brauwissenschaft*, 1972, **25**, 44.

Royal Commission on Environmental Pollution, *Lead in the Environment*, H.M.S.O., 1983.

Sherlock, J. C., Pickford, C. J. & White G. F., *Food Additives and Contaminants*, 1986, **3**, 347.

LEGISLATION AND CONTROL OF THE USE OF ADDITIVES

Brewing Research Foundation, Annual Report, *Journal Institute of Brewing*, 1987, **93**, 277.

Committee on Public Breweries, House of Commons, *Report and Minutes of Evidence*, H.M.S.O., 1818.

Committee on the Hop Industry, House of Commons, *Report and Minutes of Evidence*, H.M.S.O., 1890

Committee on Beer Materials, House of Commons, *Report and Minutes of Evidence*, H.M.S.O., 1899.

F.A.C.C., *Report on the Review of Additives and Processing Aids used in the Production of Beer*, H.M.S.O., 1978.

Leach, A. A., *The Brewer*, 1979, **65**, 129.

Leach, A. A., Long, D. E. & Ward, J., *The Brewer*, 1983, **69**, 139.

Lloyd, A., *Journal Institute of Brewing*, 1977, **83**, 68.

M.A.F.F., Food Surveillance Paper No. 14, *Progress Report of the Steering Group on Food Surveillance*, H.M.S.O., 1984.

LIQUOR TREATMENT

Rudin, A. D., *Brewers' Guardian*, 1976, **105**, 30.

Taylor, D., *Brewing and Distilling International*, 1981, **11**, 35.

MALTING ADDITIVES

Brewing Research Foundation, Annual Report, *Journal Institute of Brewing*, 1981, **87**, 126.

Davis, J. G., *Process Biochemistry*, 1972, **7**, 32.

Klopper, W. J., *Brauwissenschaft*, 1973, **26**, 109.

Narziss, L., Reicheneder, E. & Nothaft, H., *Brauwelt*, 1982, **122**, 502-519 & 627-636.

Sugget, A., Moir, M. & Seaton, J. C., *Proceedings of the E.B.C., Berlin, 1979*, 79-89.

Thalacker, R. & Kaltwasser, I., *Monatsschrift Fur Brauerei*, 1981, **34**, 115.

MYCOTOXINS

Cerutti, G., et al., *Monatsschrift fur Brauwissenchaft*, 1987, **40**, 455.

Jewers, K., *Journal of the Royal Society of Health*, 1982, **102**, 114.

NISIN

Ogden, K., *Journal Institute of Brewing*, 1986, **92**, 379.

NITRATES AND N-NITROSO COMPOUNDS

Baxter, E. D., *Ferment*, 1988, **1**, No.6, 31.

Brewing Research Foundation, Annual Report, *Journal Institute of Brewing*, 1988, **94**, 218.

Craddock, V. M., *Nature*, 1983, **306**, 638.

Department of the Environment, Pollution Paper No. 26, *Nitrate in Water*, H.M.S.O., 1986.

Johnson, P., et al., *Journal Institute of Brewing*, 1987, **93**, 319.

Johnson, P., et al., *Journal Institute of Brewing*, 1988, **94**, 19.

Kara, B. V., et al., *Proceedings of the E.B.C.*, Madrid, 1987, 663.

M.A.F.F., Food Surveillance Paper No. 20, *Nirate, Nitrite and N- Nitroso Compounds in Food*, H.M.S.O., 1987.

Mangino, A. M. & Scanlan R. A., *Journal of the American Society of Brewing Chemists*, 1982, **40**, 55.

Stone, I. M., Laschiver, C. & Saletan, L.T., *Proceedings of the American Society of Brewing Chemists*, 1968, 125.

Taylor, D. G., *The Brewer*, 1988, **74**, 532.

Wainwright, T., *Journal Institute of Brewing*, 1986, **92**, 73.

PESTICIDES

Ryman, J., *Brewing and Distilling International*, 1989, **20**, 14.

PLASTICS

Bradford, R. N., *The Brewer*, 1983, **69**, 61.

Casson, D., *The Brewer*, 1982, **68**, 447.

M.A.F.F., Food Surveillance Paper No. 2, *Survey of Vinyl Chloride Content of Polyvinyl Chloride for Food Contact and of Foods*, H.M.S.O., 1978.

M.A.F.F., Food Surveillance Paper No. 6, *Survey of Acrylonitrile and Methacrylonitrile Levels in Food Contact Materials and in Foods*, H.M.S.O., 1982.

M.A.F.F., Food Surveillance Paper No. 21, *Survey of Plasticiser Levels in Food Contact Materials and in Foods*, H.M.S.O., 1988.

PVPP

Dadic, M. & Lavallee, J. G., *Journal of American Society of Brewing Chemists*, 1983, **41**, 141.

Glynn, I. & Wilkinson, S. P., *Brewers' Guardian*, 1985, **114**, 32.

Scherrer, A., et al., *Brauerie Rundschau*, 1982, **93**, 229.

Weyh, H., *Brauwissenschaft*, 1979, **32**, 69.

REINHEITSGEBOT

Christl, J., *Brewing and Distilling International*, 1977, **7**, 49.

Ehrenstorfer, I., *Brauwelt*, 1978, **118**, 52.

Jacques, D., *The Brewer*, 1986, **72**, 420.

Mayer, K. J., *Brewing and Distilling International*, 1986, **16**, 24.

Mayer, K. J., *Brewers' Digest*, 1987, 32.

Narziss, L., *Journal Institute of Brewing*, 1984, **90**, 351.

Stevens, B., *Brewing and Distilling International*, 1981, **11**, 23.

Zeller, A., *Brauwelt*, 1985, **125**, 348.

TAINTS

Andrews, D. A., *The Brewer*, 1979, **65**, 236.

Cosser, K. B., et al., *Technical Quarterly of the Master Brewers' Association of the Americas*, 1980, **17**, 53.

Charalambous, G., et al., *Technical Quarterly of the Master Brewers' Association of the Americas*, 1974, **11**, 26.

Dadic, M., et al., *Wallerstein Laboratories Communication*, 1971, **34**, 5.

Delcour, J. A., et al., *Journal Institute of Brewing*, 1986, **92**, 591.

Schur, F. & Pfeninger, H., *Brauwissenschaff*, 1977, **30**, 134.

Voerkelius, G. A., *Brauwelt*, 1967, **107**, 1401.

WATER SUPPLY

Anon, *Which?*, February 1989.

Breach, R. A., *Ferment*, 1989, **2**, 72.

Eden, G. E., *The Brewer*, 1987, **73**, 504.

Lloyd, J. W., *Ferment*, 1988, **1**, 34.

Nicholls, G. D., *Ferment*, 1989, **2**, 66.

Taylor, D. G., *Ferment*, 1989, **2**, 76.

Taylor, D. G., *The Brewer*, 1988, **74**, 532.

GLOSSARY

Acetic acid A simple organic acid which gives vinegar its characteristic taste.

After taste A taste left in the mouth after the beer has been swallowed.

Ales Beers which are produced by a top fermentation.

Amino acids The chemicals which are the building blocks of proteins.

Amylases Enzymes which break down starch into sugars.

Ascorbic acid Vitamin C.

Attemperation Control of temperature during fermentation.

Attenuation The reduction in specific gravity of wort during fermentation caused by the conversion of sugars to alcohol.

Autolysis The process by which yeast digests itself with its own enzymes. This occurs under adverse conditions.

Barm Yeast. A word not in common use today.

Biological haze A haze caused by microbial spoilage.

Bottom fermented beers Those fermented by a yeast which settles on the bottom during fermentation. They are usually called lagers.

Brewery conditioned beer Beer which is processed in the brewery.

Brewhouse At one time synonymous with brewery but now in larger breweries it is that part in which the wort is produced.

Bright beer Beer which has been rendered clear of any suspended particles by the use of finings or filtration.

Bung hole The hole in the belly of a cask through which it is filled.

Bushel A measure of volume used for malt and barley in brewing. There are eight bushels to the quarter

Carbonation The dissolution of carbon dioxide gas in beer. It may be natural through fermentation or artificially added under pressure.

Cask conditioned beer Beer which is naturally conditioned in a cask. Real ale.

Centrifuge Equipment used for removing suspended solids in beer using centrifugal force.

Chill proofing A cold storage process by which insoluble protein/tannin material is caused to precipitate so that it can be removed from the beer.

Common brewer One who brews beer for sale.

Conditioning Any processing of the beer after the end of the primary fermentation. It may be natural as in the production of real ale or a complex series of operations carried out in the brewery.

Conditioning tank These may be warm or cold. In cold conditioning tanks a chill proofing or lagering process is carried out. In warm tanks a secondary fermentation is carried out when the beer is carbonated and flavour matures.

Conversion These are chemical changes which occur in the mash tun when more complex compounds are broken down into simpler ones. The most important chemical change is the conversion of starch to sugars.

Cooler Equipment used in the brewery to reduce the temperature of wort after boiling so that it is correct for fermentation.

Copper The vessel in which wort is boiled with the hops.

Foxing A word not in current usage but was used to describe the bacterial spoilage of wort. The smell produced was said to be similar to that of a fox.

Diacetyl A chemical produced in varying amounts during brewing for a number of reasons. It is said to have a butterscotch flavour and is found more in ales than lagers.

Dietetic beer A low carbohydrate beer in which the alcohol and the sugar content is low.

Endosperm The starchy food reserve inside the barley grain.

Enzymes Biochemical catalysts.

Essential oils Volatile flavouring compounds which give hops their aroma.

Esters These are produced by the reaction of alcohols with acids. Their fruity type of flavour is sometimes evident when they are produced in beer.

Farbe bier A specially brewed German beer with an extremely high degree of colour. It is only used in breweries, to adjust the colour of beers for sale.

Fermentation The part of the brewing process during which the fermentable sugars in the wort are converted to alcohol and carbon dioxide by yeast.

Filter media Inert substances used in filters to trap the solids in the beer which is passed through them.

First runnings The first wort to be run from the mash tun; it is the strongest.

Formaldehyde The simplest member of the group of organic chemicals known as the aldehydes. Unpleasant smell.

Glucanases Enzymes which break down glucans.

Glucans Glucose polymers which make the wort viscous and cause production problems.

Glucose A simple monosaccharide sugar. Component of other more complex sugars, starch and cellulose.

Green beer Beer from which the yeast has not been removed from suspension.

Green malt Unkilned malt.

Grist The mixture of materials from which the wort is produced.

Gyle or guile Wort which has been collected for fermentation.

Gyle tun Fermenting vessel

Hop back Vessel in which the hops and some of the trub are removed.

Hopped wort Wort which has been boiled with hops.

Humulones Bittering compounds in hops.

Hydrolysis The chemical breakdown of a large molecule into its component smaller molecules. Each break requires a molecule of water to occupy the space left on the now unattached bonds. Starch can be hydrolysed to glucose.

Isohumulones The bittering compounds found in beer which are derived from the hop.

Kilning The part of the malting process when the germinated barley is heated by passing hot air through it. This develops flavour and colour as well as drying it.

Length Total volume of a brew.

Liquor The name for water used in brewing.

Malt, brown A malt kilned at a higher temperature than pale malt. It was originally kilned with wood and used in porters when they were first produced. Not produced now.

Malt, crystal A malt in which of part of the endosperm is turned into syrup. It is then roasted to give a glassy, crystaline cut. This gives it a higher intensity of flavour and colour.

Malt, pale A light coloured malt used in brewing pale ales. They were first produced by kilning the malt with coke which did not discolour the malt as did wood.

Mashing The process by which the fermentable extract is produced from the malt and cereal brewing adjuncts.

Mashtun The vessel in which mashing is carried out.

Microflora The natural range of micro-organisms present in a particular place.

Milling The part of the brewing process in which the malt is crushed so that the liquor can come into contact with the starch inside the grain during mashing.

Modification Various chemical changes which occur in the malt grain during malting. The physical manifestation of this is an increase in friability of the endosperm.

Molasses A thick, brown, uncrystalised, bitter syrup obtained as a by-product during sugar refining.

Muscavados A low quality unrefined cane sugar separated from molasses.

Obesumbacterium proteus A wort bacteria now called *Hafnia protea*. It is the shape of a short rod and very common in pitching yeasts.

Original gravity Specific gravity of wort before fermentation.

Oxalates Salts formed during fermentation from the production of oxalic acid. Mostly calcium oxalate which is insoluble and deposited as a scale.

Pasteurisation The process by which beer is heated to a certain temperature for a prescribed length of time in order to kill spoilage micro-organisms.

pH A scale on which acidity and alkalinity are measured. Water is neutral at pH 7. Lower values indicate acidity and higher values alkalinity.

Pilsen lagers A pale type which originated in Pilsen.

Pitching yeast The yeast which is used to start a brew and is saved from one brew to another.

Polymers Large molecules which are a chain of repeated simpler compounds known as monomers. The components of starch are polymers; glucose is the monomer.

Primings Sugar added to beer after fermentation to give sweetness or to produce a secondary fermentation.

Private brewer A brewer who brewed for his own needs and not for sale. Large households usually had their own brewery before the twentieth century.

Racking The process of filling casks.

Raw grain Unmalted or unprocessed grain used as an adjunct.

Rope A condition in beer produced by bacterial spoilage. The beer develops the consistency of slime.

Running beer Beer which was sold without being matured.

Secondary fermentation A fermentation which occurs in the container (cask or bottle) in which naturally conditioned beers are stored.

Spent grains Grains left in the mash tun after all the extract has been removed.

Starch A polymer of glucose. The main form in which carbohydrates are stored in cereals.

Stock beer Strong beer which was matured before being sold.

Sulphuring The process of fumigation with sulphur dioxide. Originally carried out by burning sulphur which produces the gas.

Trub Insoluble protein/tannin material which precipitates in wort after boiling. It is also called break.

Tun A brewing vessel.

Tunning The transfer of wort or beer from the tun to the cask.

Ullage Beer left in the cask which is unfit to serve.

Wort The fermentable extract produced by mashing.

Yeast bite A harsh flavour passed onto beer from yeast when the beer is in contact with it for a longer time than is desirable.

INDEX

Picture Acknowledgements

Page 1: Bodleian Library (Douce 252, folio 34v)
Page 19: By permission of the British Library
Page 37: By permission of the British Library
Page 52: Controller of Her Majesty's Stationery Office
Page 64: Library of the Institute of Brewing
Page 111: Matthew Patton
Page 147: Library of the Institute of Brewing
Page 151: Library of the Institute of Brewing
Cover picture: An original oil painting by Rupert Gatfield